Seven Challenges for the Science of Animal Minds

Seven Challenges for the Science of Animal Minds

MIKE DACEY

Great Clarendon Street, Oxford, OX2 6DP,
United Kingdom

Oxford University Press is a department of the University of Oxford.
It furthers the University's objective of excellence in research, scholarship,
and education by publishing worldwide. Oxford is a registered trade mark of
Oxford University Press in the UK and in certain other countries

© Mike Dacey 2025

The moral rights of the author have been asserted

All rights reserved. No part of this publication may be reproduced, stored in a retrieval system, transmitted, used for text and data mining, or used for training artificial intelligence, in any form or by any means, without the prior permission in writing of Oxford University Press, or as expressly permitted by law, by licence or under terms agreed with the appropriate reprographics rights organization. Enquiries concerning reproduction outside the scope of the above should be sent to the Rights Department, Oxford University Press, at the address above.

You must not circulate this work in any other form
and you must impose this same condition on any acquirer

Published in the United States of America by Oxford University Press
198 Madison Avenue, New York, NY 10016, United States of America

British Library Cataloguing in Publication Data

Data available

Library of Congress Control Number: 2024949109

ISBN 9780198928072

DOI: 10.1093/9780198928102.001.0001

Printed and bound by
CPI Group (UK) Ltd, Croydon, CR0 4YY

The manufacturer's authorised representative in the EU for product safety is
Oxford University Press España S.A. of El Parque Empresarial San Fernando de Henares, Avenida
de Castilla, 2 – 28830 Madrid (www.oup.es/en or
product.safety@oup.com). OUP España S.A. also acts as importer into Spain
of products made by the manufacturer.

For Calla

. . . These are real questions. They are real because they all have an answer . . . Difficult cases crop up in every area of inquiry; they never give ground for general skepticism. Interspecies sympathy certainly encounters some barriers. So does sympathy between human beings. But the difficulties arising here cannot possibly mean that any attempt to reach out beyond the familiar lit circle of our own lives is doomed, delusive, or sentimental.

Mary Midgley, *Beast and Man* (1978)

© 1978 Cornell University. Used by permission of the publisher, Cornell University Press

Contents

List of Illustrations	ix
Acknowledgements	xi
Introduction	1
1. Underdetermination	15
2. Anthropomorphic Bias	46
3. Modeling	71
4. Integration and Homology	95
5. Ecological Validity	122
6. Sample Size and Generalizability	149
7. Measuring Consciousness	173
Conclusion: Of a Different Mind	197
References	211
Index	231

Illustrations

1.1	The black box	17
1.2	The competitive feeding paradigm physical layout	20
1.3	The logical problem in the black box	24
1.4	Schematic for two-stage evidentialism	31
1.5	Reframing the difference between mind-reading and behavior-reading	41
2.1	Anthropomorphism, Morgan's Canon, and the trade-off of errors	53
3.1	A causal graphical model of the Blaisdell et al. (2006) experiment	81
3.2	Example of a multilevel mechanism for spatial memory	88
4.1	The "Sloan hexagram" of interdisciplinary connections in cognitive science in 1978	96
4.2	Competing hypotheses regarding the homology of anger	106
4.3	A toy phylogenetic tree	109
4.4	Homology-based comparative claims require two steps	117
5.1	The experimental setup from Howard et al. (2019)	125
5.2	A depiction of the anchoring experiment strategy	142
6.1a	Data from Inoue and Matsuzawa (2007)	154
6.1b	Data from Inoue and Matsuzawa (2007)	155
6.2	A histogram of all experiments published in various journals in 2019	156
7.1	An example of a theory-light approach applying the multiple clusters framework	192
7.2	An example of a theory-heavy approach applying the multiple clusters framework	193

Acknowledgements

As with any book, this has been written over a long period of time, with help from many. Most of this was written at my home institution of Bates College in Maine, though a disproportionate amount was written in my stay as a Visiting Scholar at the Leverhulme Center for the Future of Intelligence at University of Cambridge in May–June 2022. Thanks to the CFI, and especially Marta Halina for making that visit possible. A number of people have contributed valuable feedback. Charles Beasley, Ali Boyle, Marta Halina, Scott Partington, and Paul Schofield all undertook the project of reading and providing feedback on complete manuscript drafts, for which I am hugely grateful. The participants in my fall 2022 Philosophy of Animal Minds seminar also read a full manuscript, catching errors and obscurities: Michelle Liu, Antonio Melgar, Andrew Monteith, Leah Moore, Adriana Pastor Almiron, and Linnea Selendy. They were a motley crew, but they really pulled together in the end (I promised them I'd write that). Thanks for feedback on chapter drafts from Jonathan Birch, Cameron Buckner, Mike Jacovides, Matthias Michel, Talia Morag, Katie Tabb, and Isaac Wiegmann. Material from the book was presented at several venues as it was developed. These include: Wishkabibble in Aurlandsdalen, Norway, in July 2019; the Northern New England Philosophical Association at College of the Holy Cross in November 2019; the Southern Society for Philosophy and Psychology, held online in December 2020; the "Conjectures and Refutations" Seminar Series at London School of Economics in November 2021; the Philosophy of Science Association in Baltimore, February 2022; the Kinds of Intelligence Works in Progress Retreat at University of Cambridge, June 2022; the Eastern Division of the APA in Montreal, January 2023; and the Philosophy of Animal Minds and Behavior Association in Madrid, April 2023. Thanks to all organizers, session chairs, participants, and others who made those possible. A few stand out in various roles from various presentations: Colin Allen, Kristin Andrews, Charles Beasley, Shereen Chang, Carl Craver, Marta Halina, Edouard Machery, Lucas Matthews, Susana Monsó, Nedah Nemati, and Jim Tabery. For helpful conversations, advice, and more general support, thanks to John Doris, Ron Mallon, and Mark Okrent. Thanks to Jen Coane and the Memory and Language Lab at Colby College for running our experiments on anthropomorphism, especially student researchers Kai Chang, Yi Feng, and Simon Xu. Thanks as well to my Bates research assistant Abe Brownell. As I was in the process of writing this, I received tenure, partly based on (or perhaps despite) a full manuscript draft. Thanks to my tenure reviewers for reading the draft and seeing fit to write supporting letters on that basis. Thanks to Bates College, especially my Philosophy colleagues Lauren Ashwell, David Cummiskey, Paul Schofield, and Susan Stark, as well as Jeanne Beliveau. Thanks to my OUP project editors Peter Momtchiloff and Jamie Mortimer as well as three anonymous readers for OUP for extremely helpful feedback. My visit to Cambridge was funded by a generous Bates Faculty Development Grant. Bruce and Vicky Wicks helpfully provided their Maine cottage for a writing retreat in August 2021. Thanks to my Mom and Dad, and to Jack, Tom, and Holly for making all this possible. Finally, thanks to Val for supporting me through the interminable process of writing a book and the buildup to tenure, even as our daughter Calla was born.

Introduction

The scientific study of animal minds is difficult. This may not be a surprising claim, but it is an important one to grapple with. We cannot look at, listen to, record, or otherwise observe a mind. Minds are the most complex objects known, and we are not sure we possess even the basic terms to describe them. Experimental psychologists can learn a lot through controlled experiments and careful recording of behavior. But results are often messy because individuals differ in how they perform a task, and any individual may change from day to day. Nonhuman animals[1] make particularly tricky experimental subjects. They don't take instructions, they can't tell us what they are thinking, and they have their own interests and purposes which often do not align with experimenters' intentions.

In addition, animal minds can often seem quite alien to us. It might be tempting to view animal minds as 'diminished' versions of our own. But we cannot understand their minds by simply subtracting from ours. Or, for that matter, by adding to ours. Any such comparison must be more complex. Animal minds relate to our minds in the same way animal limbs relate to our limbs. If we look carefully in the right ways, for example, we find that dolphin flippers share similar structure to human arms. But this shared structure may not be immediately apparent, and, more importantly for present purposes, the comparison does not itself tell us how the flipper works. We must study these traits in their own context, on their own terms. Animals live in environments that are fundamentally different from ours. They can also have fundamentally different sensory capacities (as Nagel [1974] famously highlighted in bats). Different features of the environment can have very different significance for them; they can afford very different kinds of interaction, different opportunities, and different dangers. Kuhn (1962) argued that proponents of different scientific paradigms literally see different worlds. That may go a bit far when comparing two chemists, but the point really does apply to different species. The experienced environment of the animal (the *Umwelt*, if you prefer) is likely very different from ours.

---

[1] For ease of language, I use "animals" and "nonhuman animals" interchangeably from here on out.

Seven Challenges for the Science of Animal Minds. Mike Dacey, Oxford University Press. © Mike Dacey 2025.
DOI: 10.1093/9780198928102.003.0001

2 INTRODUCTION

Despite the challenges we face, there are many reasons we should want to understand animal minds. Animal minds are interesting in and of themselves. They often display inspiring, shocking, or downright bizarre capabilities. We have pragmatic and moral reasons to want to know what conditions and behaviors are best for animals that we interact with frequently, and care for, in various contexts. Animal minds may also be interesting as reflections of ourselves in at least two ways. Nonhuman animals have long been used as models with which to study the human mind indirectly, assuming certain *similarities* (or at least evolutionary continuities). Working in the opposite direction, an accurate understanding of human uniqueness, of our *differences* from other animals, has long been considered an important part of understanding what it means to be human. Research need not focus on only one of these questions and can inform answers to any and all.

As a historical matter, the scientific study of animal minds has long lingered in the shadows of behaviorism in the anglophone world. Some versions of behaviorism assert that animals do not have minds and can be understood in simple, mechanical terms based on reinforcement and stimulus-response conditioning. This view arose in the early twentieth century and fell roughly fifty years later (at least as the dominant view in America). Since the fall of behaviorism, it has become increasingly clear that many animals have sophisticated mental lives and possess complex reasoning capabilities in various domains. I take this for granted here, and I won't argue for it specifically.[2] When we take that point seriously, we must move beyond the old, but still common, dichotomous question of whether animals are intelligent or mere machines (as if these are the mutually exclusive, exhaustive options; Halina [forthcoming]). The questions of interest become: When and how do animal reasoning processes work? Which animals have experiences, and what might those be like? Which parts of the environment are animals responding to? What information do they gather about those parts of the environment? How do they use that information to guide responses, and why do they respond that way? In more technical terms, what are the *cognitive mechanisms* driving animal behavior?

The science that has emerged to answer these questions is currently flourishing as a creative, multidisciplinary, multi-approach, theoretically diverse program. There is so much to do, and there are so many promising ways to

---

[2] My focus is on methodology rather than defending claims about minds themselves. For good general discussions of the actual abilities of animals, see, e.g., Andrews 2014, Shettleworth 2012, Halina 2024.

do it. This includes many species from various parts of the animal kingdom, viewed from a number of theoretical perspectives and using a number of methodological tools and strategies.

Some species which will play especially significant roles in the book include chimpanzees, honeybees, octopuses and cuttlefish, rats, and cats. Chimpanzees are our closest living relatives and seem, in many ways, to possess minds most like our own (see Chapters 1, 2, 4, and 6). They are a bridge to understanding animals more generally from a human perspective. Honeybees break our intuitions about what insects, which seem so simple, can do (see Chapter 5). Octopuses and cuttlefish are perhaps the most alien beings on earth (see Chapter 7). Rats (along with mice), the traditional animals of laboratory study across the biological sciences, have surprisingly sophisticated lives (see Chapters 3 and 4). Cats are relatively easy to keep in the lab because they are domesticated (see Chapter 4). This is just a small sample of species, and many of the stars of animal cognition research are left out by pragmatic necessity. A longer book might have covered work on birds such as crows, jays, and parrots, or perhaps dogs, elephants, or whales and dolphins, to name some other favorites.

Studying such wide-ranging species and differing mental abilities requires interdisciplinary collaboration, including comparative psychology, cognitive ethology, neuroscience, evolutionary biology, and philosophy. I'll discuss all of these, but philosophy and comparative psychology will be perhaps the most central. I am a philosopher by training, though I have dabbled some in the relevant empirical work. My interests and contributions here will be largely theoretical in the sense that I am primarily interested in the *interpretation* of evidence we have. Questions surrounding the interpretation of evidence are ones about which philosophers of science are well positioned to contribute significantly. Comparative psychology also falls at the center because I will focus on behavioral experiments. As a field, comparative psychology generally studies animals in the lab by performing experiments that require an animal complete some task in a tightly controlled environment. These experiments are usually seen as the most direct route to understanding the cognitive mechanisms of animals. Other fields provide crucial evidence as well. My approach, though, will be to start with comparative psychology and integrate those other findings when possible (and, indeed, questions about how to integrate across fields are central to Chapters 4 and 5).

This book addresses seven *challenges* that are faced by any attempt to interpret empirical findings into a systematic understanding of animal minds. The existence of each of these challenges is broadly recognized. I believe most

4   INTRODUCTION

researchers working on topics about animal minds are aware of all of them, at least as abstract worries. However, they have received varying degrees of explicit consideration, especially in philosophy. I do not take my list to be perfectly comprehensive, but I do intend it to capture the biggest and deepest challenges for the field. These seven challenges are, for various reasons, intrinsic to the subject of study, and they are a large part of the reason that the subject is so difficult.

I do not raise them as criticisms of existing work. My choice of prepositions is deliberate; these are challenges *for* the science, not challenges *to* the science. While framing the book around challenges may look negative or pessimistic, my orientation toward them is not. I do recognize, though, that there are features of this project that will make it hard for optimism to show through consistently. For example, this framing will lead me to spend a lot of time discussing things that we don't know about animal minds, and perhaps not enough time discussing what we do. Similarly, I will often discuss mistaken approaches to the challenges which have caused problems of various sorts. Some mistaken approaches have made the challenges appear more difficult than they are, some have biased understanding of animal minds in a particular direction (usually suppressing perceived intelligence), and some have made certain research programs appear misguided. Again, though, I take the challenges themselves to be intrinsic to the science for various reasons. Researchers have done an impressive job of working with them, around them, and through them to continue making progress.

In fact, my optimism about meeting the challenges is what makes it possible for me to write the book in the first place. I suspect, at least sometimes, that researchers shy away from meeting some of these questions head-on for fear of what they might find, like a person who suspects they are sick but avoids the doctor because they cannot bear the news. If this is right, then the failure to truly address them creates a self-feeding spiral; these issues are brought up almost exclusively as criticisms of a study. As such, it seems that this is all they can be, and to acknowledge them as legitimate is, in some important way, to concede that this research is fatally flawed. In opposition to this outlook on the state of affairs, the goal in my discussion is to spark optimism even when looking right at the challenges that the field faces.

The book is organized so that each of the seven substantive chapters addresses one challenge. The goal for each is to argue for a view of the challenge that will illuminate the way forward. Although I do not solve these problems, I present a way of thinking about them that makes them solvable. Each chapter will center on a different empirical example to illustrate the issues at hand, from mind-reading in chimpanzees to foraging in bees, to consciousness in octopuses.

The chapters are intended to be more or less modular, such that it would be fruitful to read one or two alone, to skip a chapter in a full read-through, or even to read them out of order without getting lost. There is a systematic unifying viewpoint behind my approach to all of them, and there are strong connections between the discussions. As such, a linear read-through would likely yield a fuller picture of the approach I am suggesting. But this modular structure does two main things that I think are important. First, it fronts the challenges. It is more important for the progress of the science of animal minds that it address these challenges than that it addresses them in the specific ways I propose. As such, the set of challenges deserves to be the central organizing principle of the book. Second, it provides many different 'ways in' to the material. Different literal 'ways in' mean that readers might choose to start by reading about the challenge that they find most interesting. Different conceptual 'ways in' mean that readers might find my approach more convincing in one context than in others. Despite the overarching view in the background, each challenge does require different resources, for different reasons, to address it. It's a big landscape, and I want to allow readers to find their own path through it, if they so choose.

The seven challenges, and thus the next seven chapters, are as follows.

Challenge 1: Underdetermination

The fact that experimental observations underdetermine theory is perhaps the most basic challenge here, and the one that sets up the rest. The problem is that many theoretical hypotheses can potentially fit any specific set of data obtained in experimental/observational settings. Underdetermination has long been discussed in philosophy of science as a feature of science generally; no hypothesis can be directly confirmed or falsified by a single observation. At the very least, one must evaluate the hypothesis under question along with many 'auxiliary hypotheses' that situate the target hypothesis and link it to the specific prediction. However, it has a much more pressing form in comparative psychology, where it is often the case that multiple competing hypotheses can predict (or accommodate) *either* outcome of an experiment.

This has been most often discussed in the form of the 'logical problem' in the chimpanzee mind-reading debate (Lurz 2011). The big question in this literature is whether chimpanzees attribute mental states to one another or get

6 INTRODUCTION

by with simpler 'readings' of behaviors and features of the environment. The ability to attribute mental states is generally thought to be among the most difficult cognitive tasks a creature might face, and also one with special moral significance for how we view the attributing animal. The logical problem arises because any animal can only attribute mental states *by* reading behaviors and features of the environment. So the same set of behavioral/environmental cues could drive behavior either by allowing attribution of a mental state or directly with no intervening mental state attribution. In the strongest versions of the claim, no experiment can actually tell between them (e.g., Povinelli 2020).

Challenge 2: Anthropomorphic Bias

Theorists on topics related to animal minds often worry that human beings have a tendency to anthropomorphize. That is, we tend to see human features in things that are not human, such as animals, objects, and even forces of nature. Dog owners attribute all kinds of attitudes to their pets (sometimes jokingly, but sometimes sincerely), and people tend to see faces even in simple patterns like an American electrical socket (which looks, appropriately enough, shocked). This tendency has been taken to explain religion and mythology, to ground much of human moral motivation, and has recently even been recruited to find ways to make people less lonely.

The worry for the science of animal cognition is that this tendency can bias the kinds of hypotheses that researchers find plausible. The actual implications of this tendency, and best practice to address it, have been controversial. Intentional (and thus mental) attributions pervade our normal descriptions of behavior, as when I reach *for* my coffee cup or glance *towards* the clock. So, on one extreme, anthropomorphism is taken to motivate a revision of language to completely eliminate any attribution of intention to animals. Others argue that anthropomorphism provides a valuable tool in theory construction and evaluation, though even they typically recognize that anthropomorphism can crop up in unexpected ways that might cause trouble.

The most common worry is that anthropomorphism tends to inflate the perceived intelligence of animals. As such, it has been used to motivated a principle known as Morgan's Canon, an inferential rule which dictates that hypotheses positing simpler (read: less intelligent) psychological processes ought to be preferred. Morgan's Canon, though, has been tremendously controversial, especially in recent years. In fact, I argue it is a particularly bad candidate to control the implicit, unconscious tendency to anthropomorphize. The empirical literature on *human* mind perception is the place to look to see if we can do better.

Challenge 3: Modeling

Cognitive processes are often distinguished from presumptively simpler associative processes as the two primary categories of process in animal minds. Usually, cognitive processes are thought to be more sophisticated and flexible, such as the ability to represent mental states, form maps of the environment, or reason about causal relationships. The challenge is that it is difficult to model these processes with any precision. Usually, as a result, the hypothesis is merely expressed verbally: "chimpanzees attribute mental states to one another" or "rats are capable of causal reasoning." The details of these claims might be filled in with further explanation. However, it is never fully specified exactly when or how such a capacity is engaged. This is not the fault of theorists; it is plainly difficult to describe psychological processes, in part because we lack good tools for doing so.

This has caused significant problems. Arguably, the lack of specificity of cognitive models has contributed to underdetermination (Challenge 1). The fact that underspecified cognitive models are often filled in by folk psychology also, arguably, exacerbates anthropomorphic bias (Challenge 2). Finally, the consistent role of associative models (that simply link mental representation of features of the environment using associations) as the contrast to cognitive models has produced a large divide in the space of options: cognitive hypotheses are typically quite sophisticated, while associative hypotheses are usually seen as extremely simple (e.g., Buckner 2017). Realistically, many animals operate in the space between these. The lack of articulated models has also arguably led to strategies of "trophy hunting," where researchers simply attempt to show exciting effects rather than test models at all (Allen 2014).

Challenge 4: Integration and Homology

While I mostly discuss comparative psychology here, focusing on behavioral experiments, the science of animal cognition is a distinctly interdisciplinary project. Despite the lofty goals of integration in cognitive science more

8   INTRODUCTION

generally, and real recent progress on this front, it has not gone as smoothly as one might wish. Different disciplines ask different questions, target different parts of the system, and apply different methods. It is, thus, a generally difficult problem to articulate how we can actually bridge disciplinary gaps and allow ideas and findings from different disciplines to meaningfully interact.

More specifically, inferences from evolution and neuroscience to cognition very often rely on claims (or assumptions) of *homology*: that two traits in different species are derived from a single trait in the most recent common ancestor. This is the relevant sense in which traits are considered to be 'the same' in ways that support inferences about the trait in one (target) species based on what we know about that trait in another (source) species. Homology is thus required for most of the truly *comparative* claims in comparative cognition.

However, establishing cognitive, or functional, homologies across biological taxa presents its own challenge. This is true even in the case of work on the basic emotions, which is probably the best candidate for integration in the cognitive sciences; the basic emotions seem the most likely candidates to be evolved and directly wired into neural networks. However, as a general matter, claims of homology in psychology face two related problems. First is a *boundary problem*, where it can be unclear how to actually delineate candidate homologies to one capacity as opposed to another. For instance, is *this* observed behavior in cats a homologue to human anger, human fear, or something else entirely? Second is a *specificity trade-off*. If we make our homology claims very vague, we might be able to establish that there is likely a homology, but it can be hard to know what to conclude from it. If, instead, we make them very specific, we can potentially say a lot, but it is usually less likely that the homology claim is true. Attempts at integration often simply assume homology, and thus fail to address the boundary problem, or fall too far on either side of the specificity trade-off.

Challenge 5: Ecological Validity

Laboratory experiments are often criticized for a failure to be ecologically valid. The worry is that the artificiality of the experiment makes interpretation difficult and generalization outside the lab impossible. There are a number of ways an experiment may fail to be ecologically valid. The laboratory context itself is artificial, as often are the materials used in an experiment. The task asked of the animal may be artificial or, at least, may have little connection to the natural roles of experimental stimuli and context. As such, experimental results obtained in a lab may be misleading; the animal may be confused or distracted by the artificial environment, or it may not interpret the task in the way that the experimenter intends. Even if these problems are avoided, a laboratory experiment may tell us nothing about the actual lives of members of the species. At the same time, laboratory experiments remain essential because they allow precise control that is impossible in field observations.

Thus, it can be difficult to be confident about what we ought to infer from any experimental result. Even to those most optimistic, it may be tempting to think that different branches of comparative cognition are pursuing entirely distinct goals. Laboratory studies are interested in the cognitive mechanisms present, with little interest in the animals' lives, while field observations are interested in the animals' lives with little interest in cognitive mechanisms. But if we are to fully understand animal cognition, we need to be able to bridge these and work past the interpretive problems that come with artificiality. Work on honeybee foraging offers especially promising examples of how this might go.

Challenge 6: Sample Size and Generalizability

A number of practical constraints limit the number of individual animals that can be included in most laboratory studies. As a result, studies have been frequently performed with single-digit samples of participants, and it has not been uncommon to have only one. My main example here will be a study on chimpanzee working memory that draws conclusions that seem to rely on the performance of a single chimpanzee. With such small numbers of individuals involved, it is difficult to tell what conclusions can be generalized from those individuals to the species as a whole.

Recently, attempts to replicate many experiments in other parts of psychology have failed at disturbingly high rates. The resulting 'replication crisis' has motivated reconsideration of basic statistical practice, including sample sizes. As a result, researchers have begun to find ways to increase sample sizes. I encourage these, but they are unlikely to completely address the problem. Moreover, even if samples increase in future studies, we might still want to interpret already existing experiments. What, if any, interpretive value is provided by these extreme small sample studies? If we take them to provide the same sort of evidence as any other experiment (as has been practice), how do we address worries about generalizability?

10 INTRODUCTION

Challenge 7: Measuring Consciousness

Consciousness, or the bare capacity for having experiences, is one of the most challenging subjects imaginable for scientific study. It is intrinsically private and unobservable from the outside. When we want to know what human participants are conscious of, we ask them. This is not an option for non-human animals. It is also not even clear what causal role consciousness plays in behavior, as it seems perfectly possible that an unconscious automaton (or zombie) could be programmed to behave the same way. Even worse, it is difficult to even understand how a first-person, conscious perspective is possible in a physical universe. Influential philosophical work has stressed various ways that consciousness does not seem to fit in the standard objective, mechanical scientific view of the world (Nagel 1974; Levine 1983; Chalmers 1995). For all these reasons, scientists and philosophers alike have long tended to avoid discussion of consciousness in other species, since it seems impossible to know how to make concrete progress.

More specifically, if we want to know which species are conscious, we need indicators or measures of consciousness. Some candidates might be based on substantial theories of consciousness. Others might be based on thinner assumptions (for example, the thought that conscious processing facilitates some behavior, or that behaviors motivated by pain will aim at the reduction of pain). However, both types of measure, each in a different way, face two core problems. They are highly uncertain as measures, and they can't tell us what to make of *edge cases* where observations don't cleanly fit as indications of sentience or its absence.

Perhaps the most intriguing case, here, is that of octopuses, which display sophisticated behavior, but with whom we do not share an ancestor that had a brain. So any direct comparisons here seem impossible. Even so, consciousness, especially the capacity to experience pain or pleasure, is of tremendous importance in moral decision-making. The capacity to suffer is often (reasonably enough) viewed as a precursor to being treated as a legitimate subject of moral concern.

For each of these challenges, I suggest specific ways of reframing the problem that I think will be more fruitful. I don't claim to solve them, but to point in a direction where I think solutions are more likely to be found. The overall view that emerges in these specific discussions is one in which evidence must be gathered, organized, and evaluated holistically. This involves *modesty* about how much we learn from any one 'piece' of the scientific project: any individual behavioral experiment, any individual appeal to background neuroscience or evolutionary theory, or any individual model of a psychological process. Each is severely limited in ways that are not always appreciated. The reframing of each challenge amounts to suggesting specific ways of thinking about experimental findings, models, and inferences that help ensure the right sort of modesty. These include practices in experimental design, cross-disciplinary inference, modeling, and model choice that help get past the challenges. Overall, the evaluation of hypotheses about animal minds requires a holistic inference to the best explanation based on various kinds of evidence with various scaffolds and guardrails built in.

The scientific project that would result is a difficult one. The evidential value of experiments, observations, and features of background theory must be argued for, and can no longer be assumed. This opens up lots of room for argument and disagreement. It requires judgment calls and individual assessments of reasonableness. The holistic consideration of many pieces of evidence requires that all these different kinds of evidence be gathered, and opens many possible points of disagreement. It might strike some readers that it is implausible that the field could reach a consensus in such a project, where so many factors might influence evaluation of models and might be assessed differently by reasonable people. This difficulty, however, is intrinsic to the subject matter itself. We face these seven challenges whether we like it or not, and there are no easy answers. We cannot hide behind simple rules or rigorous-seeming but misapplied statistics. The approach here at least brings the disagreements into the open and gives researchers common ground on which to discuss them.

This approach contrasts with one common approach to the science of animal minds, which I'll call "standard practice" (following Buckner 2011). Standard practice, as I see it, is built on the conjunction of four features:

a. **Morgan's Canon:** The general preference for models that describe simpler psychological processes.
b. **Associative/cognitive divide:** Treating the choice between competing hypotheses as a choice between one option which is associative and one cognitive, with a large gap in perceived sophistication between them.
c. **Success testing:** The framing of most experimental designs around the question of whether an animal 'succeeds' at some task.
d. **Null hypothesis significance testing:** A statistical method that frames experimental statistics around the rejection of a 'null' hypothesis.

12 INTRODUCTION

Philosophy of animal minds has been the location of rich and thoughtful discussion of methodological issues. Much of this work has challenged each of these features of standard practice. For example, much discussion has focused on Morgan's Canon, which dictates that researchers prefer explanations that posit simpler psychological capacities (e.g., Sober 1998; Karin-D'Arcy 2005; Fitzpatrick 2008, 2017; Andrews and Huss 2014; Meketa 2014; Dacey 2016a). This rule has arguably dominated strategies for evidential evaluation in comparative psychology and has thus structured the discipline at large, including modeling and experimental design. There have been several justifications offered for Morgan's Canon, from worries about anthropomorphism to theoretical conservatism, to appeals to evolutionary biology. It has also received substantial criticism, arguing that its attempted justifications fail and that it has a pernicious effect by suppressing the perceived intelligence of animals and shutting off potential research programs.

The associative/cognitive divide arises from a distinction in types of psychological processes that puts them in two broad camps, either *associative processes* or *cognitive processes*. Associative processes are generally thought to be very simple, automatic, inflexible, and even mechanical. Cognitive processes are often set up to be very sophisticated, perhaps modeled on human capacities. Arguably, this effectively recreates the behaviorist framing of the science, despite the decline of behaviorism as an explicit stance. It effectively frames much of the field around the question "are animals intelligent or are they mere machines?" This framing, coupled with Morgan's Canon, has arguably distorted the project in various ways. Very probably, many animal minds fall somewhere between the very different options (or perhaps they ought to be measured on an entirely different scale or multidimensional space). Halina (2022) argues persuasively that there are lots of ways of being intelligent, so we ought not to assume that 'intelligence' is a single thing. The specific role of the concept of association has also been a target of criticism, including my own (Buckner 2011, 2017; Dacey 2016b, 2017b, 2019a, 2019b).

Recently, Taylor et al. (2022) have criticized success testing. They advocate a move away from the common experimental design practice that tests whether animals "succeed" in a certain task and a move towards testing for "signatures" of capacities instead ("signature testing"). This requires a change in the way that experiments are designed and, as a result, the direction research programs would take. The more specific statistical method of null hypothesis significance testing has also come up for criticism, especially in light of recent failures in attempted replications of experiments in other branches of psychology, which have come to be known as the 'replication crisis' (Ebersole et al. 2016; Klein et al. 2014; Open Science Collaboration 2012).[3] The discussions include basic statistical methods (Machery 2021; Wagenmakers et al. 2011) and evaluation of experimental evidence more broadly (Yarkoni 2022; Diener et al. 2022; Lin, Werner, and Inzlicht 2021; in comparative psychology see the special edition of the journal *Animal Behavior and Cognition* 8(2) [2021]).

I take these existing criticisms to identify real problems, and many of them to be broadly pointing in the right direction for solutions. Thus, the book aims to push existing discussions forward in a few ways. Some of the seven challenges are well known and much discussed, in which case I argue for a reframing of existing discussions. Other challenges have only (or mostly) been discussed in negative terms, in which case I suggest more positive directions. Still other challenges are unrecognized or ignored in the philosophical literature, in which case I attempt to frame them and sketch a means to address them.

I also think it is important to look at all the challenges together, collectively, as in this book. Existing discussions, as insightful and valuable as they are, have typically addressed these challenges in a piecemeal manner. But actually addressing the challenges to the science, in light of concerns about various practices, methods, and ways of framing the field, requires a systematic, positive alternative picture of the science. The four features of standard practice fit together so well and operate so smoothly that simply criticizing one at a time won't be enough to replace the entire structure. As a general matter, it is helpful to be able to offer an alternative to attract adherents away from the old view. I'd even suggest, these specific four features are so strongly mutually supporting that even if one is removed, the others will effectively replace it; a systematic replacement is necessary to truly move on.[4] But I know of no alternative account of the science that approaches the systematicity of standard practice.

As I work through the seven challenges in this book, a more systematic alternative emerges. This replacement arguably amounts to a substantial change to some common ways of viewing and evaluating evidence, interpreting (at least some) models, and some ways experiments are designed and chosen. Nevertheless, I recognize (and hope) there are plenty of fellow travelers already out there. The methodological discussions just mentioned often point in a similar direction, and researchers in various parts of the field may be traveling similar paths already. To them, I hope it is helpful to consider a bigger-picture

---

[3] Though I focus on comparative psychology, I think all of the challenges discussed here apply across psychology, in different ways in different subfields. Likely, some are less pressing in parts of human psychology for various reasons.

[4] I'll defend this claim in the Conclusion.

sketch of the way. To others, parts of standard practice may seem so entrenched that an alternative is hard to imagine; in which case, researches may agree with critical arguments but may remain unable to actually incorporate them in research. I hope a sketch of the positive path can help here as well. I can only provide a rough sketch for such an overarching view, but that sketch is, I think, an important step.

The features of standard practice will come up at various points through the book, for different reasons and to different ends in different chapters. I'll summarize the positive proposals more systematically in the Conclusion, when all the pieces are in place. Though I do not organize the book around this overarching view, the sustained attention to all seven challenges allows the central themes to emerge. Moreover, to my own mind, a focus on the actual challenges the science faces is the best way to argue for a view of how the science can proceed.

While these seven challenges remain daunting, and the research approach I sketch is difficult, we have come a long way in the relatively short time that the science of animal cognition has been pursued in earnest. We can, and will, continue to make progress. The strategies I outline here are not the only way forward, but I believe that they will help.

1

Underdetermination

1.1 Introduction

Premack and Woodruff (1978) initiated one of the most active and contentious literatures in comparative psychology when they asked, "Does the chimpanzee have a theory of mind?" They approached the question in a series of experiments with a chimpanzee named Sarah. In one of the basic versions, Sarah was shown a 30-second video of a human actor attempting to accomplish a task—for example, attempting to reach a banana that was stuck to the ceiling, out of reach. At the end of the video, Sarah was offered two photographs, one of which depicted a solution to the problem (in this case, stepping onto a stool), and one of which did not. She reliably chose the photograph with the solution. In subsequent experiments, they varied the actors as well as the problems faced and the kinds of solutions displayed. They conclude that Sarah consistently chose the right image and argue that this supports the claim that she has a theory of mind. In short, the explanation is that she recognizes that the actors in the video have a *goal* in mind and *beliefs* about how to accomplish it despite a depicted problem.

This interpretation, however, was challenged in peer commentary published with the landmark study. Most influentially, Dennett (1978) argued that, in order to really demonstrate that a participant like Sarah is attributing beliefs, we would need to find evidence of the attribution of *false beliefs*. That is, we can't tell whether Sarah is attributing true beliefs to a character in the pictures, or whether she is acting on her own beliefs about the situation. If attributed beliefs are false, though, expectations based on the attributed belief and expectations based on the world come apart (thus was born the idea of the "false belief task"; more below). Overall, the critics of Premack and Woodruff's interpretation have won out; this work is remembered more for the question it asked than the answer it gave.

This exchange, where a result initially purported to support some model of animal minds is challenged with alternative explanations, has been typical in comparative psychology. The problem is that the data *underdetermines*

Seven Challenges for the Science of Animal Minds. Mike Dacey, Oxford University Press. © Mike Dacey 2025.
DOI: 10.1093/9780198928102.003.0002

theory; many theoretical claims can be consistent with existing data. Underdetermination is an old and well-known problem in philosophy of science, sometimes framed around the Quine–Duhem thesis: at a minimum, a number of 'auxiliary hypotheses' (background beliefs about the world or experimental context) are required to connect any observation to a target hypothesis that is under evaluation. Even so, there usually remain many logically consistent hypotheses that could predict the observation (Quine 1951). Sciences don't have to worry about this as a practical matter as long as there is a limited set of hypotheses on the table and general agreement on how a given experiment bears on their evaluation. However, this pragmatic escape is often unavailable in comparative psychology. As a result, the field faces a particularly trenchant version of underdetermination.

Minds themselves are not observable. So, if we want to understand them, we must make inferences from observable phenomena to claims about the nature and operations of the mind. Typically, in comparative psychology, the observations will be behavioral responses of animals to a particular situation or task. The challenge of underdetermination arises because this inferential step from behavioral observations to theory is uncertain: many unobserved mental processes could produce the observed behaviors.

The debate over whether chimpanzees attribute mental states to one another, which has unfolded since Premack and Woodruff's (1978) original paper, will be my primary empirical example in this chapter. As human beings, we reason about what others are thinking by observing their behavior. The examples can be mundane, but the domains, cues, and responses vary: I can see you glance towards the last slice of pizza and guess that you might want it (so I should move quickly); I wonder how you knew the punchline to my joke before I told it; I guess your plan in our board game and change my own in response; I recognize a look of confusion in your face as I attempt my explanation, so I try one more example; as I approach a friend, I see from his posture that he looks distraught. This ability has come to be known as "mind-reading" (which I will use over "theory of mind," though these are often used interchangeably). The debate about whether chimpanzees mind-read has gotten significant attention in recent decades, including in philosophy. This is the case in part because of the apparent significance of the capacity itself. It has been traditionally viewed as a strong candidate to mark human uniqueness, and seems a central part of our social and moral psychology. But it is also because this debate has run up against a particularly trenchant version of underdetermination known as "the logical problem." It is, as a result, the paradigm example of underdetermination in comparative psychology.

1.2 Underdetermination and the Logical Problem

This chapter proceeds in three main sections. In Section 1.2, I present what I take to be a reasonably uncontroversial summary of the logical problem in the chimpanzee mind-reading debate.[1]

1.2.1 Inferring into the Black Box

In many ways, the underdetermination challenge sets the background for the rest of the challenges discussed in this book. That is, many of those challenges are difficulties that arise as we attempt to bridge the inferential gap from behavioral evidence to theories about the mind. To adopt a common characterization, the mind of another animal is a black box. We can only observe the inputs and outputs, but not what happens inside (see Figure 1.1). The job of the psychologist is to determine what machinery is operating inside the box, linking those inputs with those outputs: what information is encoded, how is it represented, and how do those representations cause responses? This inference *into* the box is where underdetermination arises.

Often, perhaps most of the time, experiments are designed such that the key question is whether an animal can *succeed* in some task ('success testing' in the Introduction[2]). The inputs, then, would be different variants of a task (or varying context, materials, or stimuli), and the outputs would be behaviors that amount to success or failure on the task. So, an animal may be presented with a task presumed to require a capacity to reason about the mental states of others, and successful completion of the task is taken to show that the animal may have that capacity.

That apparently simple inference is complicated by practical and theoretical problems. On the practical side, animals may not be motivated to perform

Fig. 1.1 The black box.

---

[1] Those well versed in this literature might find that things start to get more interesting in Section 1.3.
[2] Taylor et al. (2022) criticize this approach. More below.

18 UNDERDETERMINATION

a task, or they may not interpret the task as the experimenter intends. As a result, they may 'fail' merely out of lack of interest, or they may 'succeed' by some unanticipated means other than the target capacity. So success does not necessarily indicate the presence of the capacity, nor does failure indicate its absence. Put another way: minds are complicated, and there are any number of processes running at a time which may drive behavior. Individual animals also differ in their mental makeup (across subjects), and any individual may perform or interpret a task differently across time as motivation and moods shift (within subjects). They might also interpret the task or context differently from what experimenters intend; they have different sensory abilities or needs, live in very different environments, and have different ways of interacting with the world. So they might pick up different cues or attribute to them different significance. We can limit these issues somewhat in human experiments by telling participants exactly what we want them to do, asking them how they interpret the task, and allowing them to tell us when they are uninterested. These checks are obviously unavailable with animals. One hopes that repeated trials and large samples of individuals can address some transient problems with motivation or individual differences. But even so, statistical variance is characteristically quite high in the data gathered in behavioral experiments, and effect sizes are often quite small. These practical issues can make it difficult to gather decisive evidence for one hypothesis about the mind over another.

On the theoretical side, formulating and differentiating models of animal cognition can be extremely difficult. Animal minds can be quite different from our own. As such, it can be difficult to generate or comprehend detailed hypotheses about them. It's likely that many or most of our first-pass hypotheses simply don't map onto the processes that operate in animal minds. This means that we often find results that seem to provide preliminary support for a particular hypothesis, only for later results in other tasks/contexts to show that that hypothesis could not fully capture the process(es) responsible. Finally, and perhaps most troublingly, competing hypotheses can make the same prediction in a given experiment. In some cases, this is because the hypotheses have overlapping and interdependent components. To the extent that this holds, it is not clear that a behavioral experiment can decide between them at all. Some have argued that the chimpanzee mind-reading debate is in just such a position.

We saw some hints of this in the reaction to the Premack and Woodruff (1978) experiments that opened the chapter. I'll now run through three more

UNDERDETERMINATION AND THE LOGICAL PROBLEM   19

landmark experiments in the literature to show the challenges of experimental interpretation more fully.[3]

1.2.2 Landmarks in the Chimpanzee Mind-reading Debate

The next wave of significant experimental work on chimpanzee mind-reading came with work by Povinelli and colleagues in the early 1990s. Povinelli, Nelson, and Boysen (1990) tested whether chimpanzees differentiated between a human confederate who *knew* where food was hidden and one who was simply *guessing*. In these experiments, there were two cups and two human experimenters who each indicated a cup to the chimpanzee (the chimpanzees could not see where the food was hidden). In the first phases, the chimpanzees were able to observe one of the human experimenters hiding the food (the 'knower'), while the other experimenter left the room and returned after the food was hidden (the 'guesser'). All four chimpanzees reliably selected the cup indicated by the knower over the guesser. In a subsequent transfer stage, the guesser remained in the room but covered their head with a paper bag for the actual hiding. Three of the four chimpanzees were able to transfer to the new condition. The authors note that chimpanzees are likely learning rules along the lines of "select the person who hides the food" but still take the experiments, overall, to suggest that chimpanzees not only are capable of visual perspective taking but understand that perceptual access means that others will have knowledge of the situation.

However, subsequent reanalysis and further experimentation led Povinelli to doubt this conclusion, and he ultimately became perhaps the most influential skeptic of chimpanzee mind-reading (e.g., Povinelli and Vonk 2003; Povinelli 2020). For instance, individuals in the transfer test performed at chance in the first few trials (Heyes 1993, Povinelli 1994). So, perhaps they are relearning rather than transferring knowledge from the previous task. Povinelli and Eddy (1996) created a task variant based on the chimpanzees' natural begging behaviors: reaching out with palms upturned to ask for food. They reasoned that this had the benefit of being a natural response, where the previous experiments were more artificial and required more training. They

---

[3] This selection is meant to be illustrative rather than comprehensive; there are certainly other experiments that might be taken to provide the best evidence of mind-reading.

found that chimpanzees preferentially begged towards experimenters that were facing towards them rather than away, but they did not preferentially beg towards experimenters who were looking down at them rather than up at the ceiling or towards experimenters who had blindfolds over their mouths rather than over their eyes. Thus, Povinelli and Eddy conclude that chimpanzees respond to body orientation rather than eye direction, weakening support for the mind-reading explanation. Povinelli has since argued that there is no evidence that chimpanzees, or any other nonhuman animals, "possess anything remotely resembling a 'theory of mind'" (Penn and Povinelli 2007a).

Perhaps the most frequently cited evidence for chimpanzee mind-reading comes from a set of experiments by Hare and colleagues (Hare et al. 2000; Hare, Call, and Tomasello 2001), which pitted a subordinate chimpanzee against a dominant in competition for food. Generally, in a direct competition over food, a dominant will win. In this experiment, subordinates and dominants were positioned on opposite sides of a small room, each with a door they could see through. In some trials, the subordinates were allowed to observe the placement of two morsels of food, while the dominants were not. One of the morsels was placed behind a barrier, such that the dominant could not see it, while the other was out in the open (depicted in Figure 1.2). In these trials, the subordinate could get to that morsel and eat it while the dominant was occupied with the food it saw first. These authors found that the subordinates were more likely to first go after the hidden food, perhaps suggesting that they understood what the dominant could and could not see, and how they could exploit it. In other variants, the subordinates seemed sensitive to

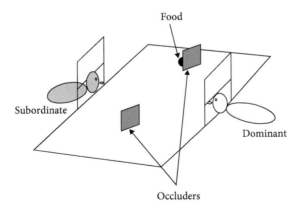

Fig. 1.2 The competitive feeding paradigm physical layout. Reprinted from Brian Hare, Josep Call, and Michael Tomasello, "Do chimpanzees know what conspecifics know?" Animal Behaviour, 61(1), 139–151, © 2001, with permission from Elsevier.

whether a dominant had been present to observe the placement of a hidden piece of food. They also preferentially went for hidden food when given a head start, indicating that they were not merely reacting to the movements of the dominant.

Though this set of experiments has been influential, it has also been criticized because it does not rule out the possibility that the subordinate is representing a 'direct line of gaze' between the dominant and food, rather than representing the dominant as seeing the food (Lurz 2011). That is, the subordinate may recognize certain physical features of the layout of the environment relative to the dominant such that there is a direct line from the dominant's eyes (or face, or head) to one piece of food and not another. It could be the case, the argument goes, that this allows a successful response without actually attributing a mental state like "seeing" to the dominant. Geometric orientation is thought to be easier to extract and represent from the visual environment than abstract unobservables such as mental states.

Most recently, Krupenye et al. (2016) purport to have demonstrated representation of false beliefs in great apes (including chimpanzees, along with bonobos and orangutans). In this experiment, apes were shown a video while cameras tracked their gaze across the screen. These videos depict a scene with two 'haystacks' in the foreground and a door in the background. In familiarization videos, an actor in an ape costume ('King Kong') hits a plainly dressed human actor. The actor runs into a door and emerges with a rod. At this point, King Kong hides inside one of the haystacks, which the human hits with a rod. In the false belief variants, the human observes King Kong hiding in one of the stacks before leaving to get the rod. Now, though, King Kong exits the haystack and runs offscreen (briefly visiting the other haystack first), while the human is absent. Krupenye et al. tracked the gaze of the apes as the human then came out of the door and approached the foreground. Of the apes that looked at the haystacks, most made their initial glances at the stack in which King Kong had hidden while the human was present, despite the fact that King Kong was absent. They take this to show sensitivity to false beliefs: the ape subjects know that the human wants to hit the stack with King Kong in it, and they know that the human falsely believes King Kong is in the stack that he saw King Kong hide in. Subsequent follow-ups appear to support claims that the effect here is a reaction to the presence of another social agent and not asocial cues (Kano et al. 2017), and that apes are sensitive to whether agents have visual access, as provided or prevented by transparent and opaque barriers (Kano et al. 2019).

False belief tasks have been a gold standard for mind-reading in the literature on human development since they were proposed in response to Premack and

22 UNDERDETERMINATION

Woodruff's original experiments (Dennet 1978). The idea is that false belief tasks require the ability to recognize that some other individual's beliefs might differ from reality. So, for example, if a participant demonstrates the expectation that another individual will act according to the world as it is, that participant may simply be representing the world itself. However, if an experiment can demonstrate that participants expect another individual to act according to beliefs about the world that they know to be false, we can be confident that participants are actually attributing beliefs. The classic false belief task in developmental psychology simply asks young children what they expect, though there have been many variants (Wellman, Cross, and Watson 2001). Despite the pedigree of false belief experiments, again, this experiment arguably fails to rule out non-mind-reading hypotheses. For instance, Heyes (2017) suggests that the apes are simply looking at the last haystack that the human interacted with before leaving the scene.[4]

1.2.3 The Logical Problem

The big question in the mind-reading debate is whether or not chimpanzees possess the ability to attribute mental states to others. The debate typically pits two *umbrella hypotheses* against one another. One hypothesis is that chimpanzees do, at least sometimes, attribute mental states. This is the mind-reading hypothesis. The other is that they do not, and instead get by with a set of capacities that work without attributing mental states, often called "behavior-reading," because it includes making predictions based on behavior rather than mental state attributions. I refer to these as 'umbrella hypotheses' to differentiate them from the more specific models we might categorize under either heading. This also separates the big question from any smaller questions of interest that might arise.

The pattern just outlined is one in which each finding, initially taken to demonstrate a mind-reading ability, was immediately challenged with plausible alternative explanations. Admittedly, the impact of these challenges is not uniform. The field seems largely convinced by challenges to Premack and Woodruff's picture-matching paradigm, as well as Povinelli's knower/guesser

---

[4] The Kano et al. (2017) experiment above is intended to address this concern by replacing a video of agents with a cartoon in which simple, solid shapes interact. Heyes evidently consulted on this test; however, it seems to me that it suggests that apes see this as a social task of some sort but not necessarily one of belief attribution. So, the results could still be explained by a form of social cognition that doesn't involve actual mind-reading.

paradigm (or, at least, this is not the ground on which proponents have chosen to fight). Interpretations are split on Hare et al.'s competitive feeding paradigm, and the response to Krupenye et al.'s false belief variant to date suggests the same.

Much of the skepticism over mind-reading interpretations of these experiments is driven by the mundane kind of underdetermination I've discussed so far. Minds are complicated, and there are many kinds of process that might be responding to many different features of the environment in many different ways. We cannot see what they are doing, so we must infer based on the patterns we can observe. It's also worth noting that the samples of animals involved are typically small (the experiments discussed above use, in order, one chimpanzee, four chimpanzees, twelve chimpanzees, and forty great apes[5]), and the response patterns are far from uniform. For example, in the Krupenye et al. false belief experiment, only thirty of the forty subjects even looked at one of the two haystacks, and twenty of those thirty looked at the 'false belief' stack first. In the competitive feeding experiments in which subordinates got a head start, they took their first step towards the hidden food 73.4 percent of the time (Hare, Call, Agnetta, and Tomasello 2000, 780). These are effects, to be sure, and this kind of variance is not unexpected. But when skeptics already have concerns about the experiments, this messiness can challenge interpretation.

The mind-reading debate, in addition, has faced a particularly challenging version of underdetermination known as the logical problem (Lurz 2009). The logical problem arises from what Povinelli (2020) calls the 'asymmetric dependency' of mind-reading abilities upon behavior-reading abilities. The idea is that any posited mind-reading capacity will depend on a prior behavior-reading capacity in order to work. That is, any chimpanzee attributing a mental state to a conspecific would *have to* do so on the basis of cues in the directly observable behavior and environment of the conspecific. This may include things like the orientation and posture of the conspecific, its location and the geometry of the space, perhaps facial expressions or other movements. Any mind-reading hypothesis will have to specify which cues the chimpanzee is using. The logical problem arises because it is always possible to construct a *complementary behavior-reading hypothesis* that says the chimpanzee is directly responding to those cues without forming an intervening representation of a mental state (Lurz 2011; I depict this in Figure 1.3).[6] Much of the debate

---

[5] Specifically, Krupenye et al. (2016) included fourteen bonobos, nineteen chimpanzees, and seven orangutans in the particular experiment described here.

[6] Thus, the problem is 'logical' in the sense that it arises out of the very logic of attempting to differentiate mind-reading from behavior-reading based on the cues that an ape responds to in any given experiment.

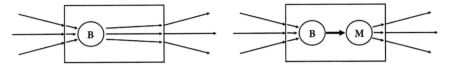

Fig. 1.3 The logical problem in the black box. 'B' indicates a representation of behaviors of another agent, while 'M' indicates a representation of a mental state in another. Fig. 1.3a diagrams a behavior-reading capacity, and 1.3b diagrams mind-reading. The logical problem arises because the observed inputs and outputs (into and out of the box) can be identical in a given experiment.

on chimpanzee mind-reading has been animated by the logical problem. I'll highlight two specific features of the debate that demonstrate the centrality of the logical problem.

First, in the absence of decisive empirical evidence, parsimony (simplicity) arguments have historically played a significant role in arguments for each side. For example, mind-reading proponents have argued that their view is more parsimonious because a single mind-reading capacity could explain responses across a number of experimental contexts, where, supposedly, several behavior-reading explanations would be required. For example, Tomasello and Call (2006) argue in favor of the mind-reading hypothesis, calling themselves the 'boosters':

> parsimony is often raised as an issue . . . it would seem that the boosters clearly have parsimony on their side. The number of different explanations required to explain the evidence available is sensibly smaller. (380)

In contrast, skeptics have argued that their model is simpler on the grounds that it posits simpler processing capacities (an application of Morgan's Canon), and that it requires fewer processing stages and representations. For example, Penn and Povinelli (2007a) argue:

> The suggestion that mentalistic explanations are more 'parsimonious' than non-mentalistic explanations is a constant refrain in the comparative literature. Our . . . hypothesis provides a representational-level explanation for why the putative 'parsimony' of mentalistic explanations is illusory . . . Positing that nonhuman animals are limited to reasoning about first-order relations between observable states of affairs is thus both more consistent with the comparative evidence and more parsimonious from a computational and representational point of view. (76–7)

These parsimony claims treat different features of the target systems as simple in the desired way: is it a matter of counting the number of individual processes one needs to posit, or counting the number of representations used in a process, or is it the simplicity of the process itself? Because they appeal to different things, these parsimony claims tend to simply clash rather than resolve the issue (I'll expand on this below; see Table 1.1 below for a summary of the varieties of parsimony claim mentioned in this chapter). Each side champions their own version of parsimony and dismisses their opponents. Which argument we take to be convincing depends on which kinds of parsimony we like most, not on the substance of the experiments (and, as such, parsimony claims may often be chosen ad hoc to defend one's preferred hypothesis). I'll say more about this below, but for now the point is simply that appeals to parsimony are common and make up a significant part of arguments for each side because the evidence itself has not been convincing across camps.

Second, a considerable amount of time and energy in the literature on chimpanzee mind-reading has been dedicated (both by philosophers and psychologists) to proposing experiments that, *if they could be done*, would finally avoid the logical problem and settle the matter. This has been a consistent feature of the discussion in recent decades (e.g., Heyes 1998; Povinelli and Vonk 2003; Penn and Povinelli 2007a; Lurz 2009, 2011; Buckner 2014; Sober 2015). And yet, these proposed experiments have generally either proved impossible to run or returned the result that failed to settle the debate.[7]

These authors generally disavow the claim that this is a hunt for a single crucial experiment. As Lurz puts it, citing the general problem of underdetermination in philosophy of science:

> the logical problem is not a search for a "decisive" test, for there are no such tests in empirical science . . . solving the logical problem, rather, involves designing a test (or tests) the positive results of which are more plausibly explained or predicted by the mindreading hypothesis than by its complementary behavior-reading hypothesis or the other way around. (2011, 29–30)

This point is well taken. But the very idea that we just need to find the right experiment has acted in the literature, for practical purposes, just like a requirement for a crucial experiment. Indeed, the two features of the debate just

---

[7] Lurz, for his part, was convinced by a positive result in his suggested experiment (involving mirrors), arguing that it does provide convincing evidence of mind-reading, though this is not uncontroversial (Lurz and Andreassi 2021).

26 UNDERDETERMINATION

described reveal two interrelated assumptions that I argue bring us too close to a critical experiment framing.

1.3 Revaluing Evidence

In Section 1.3, I shift to my own reading of the mind-reading debate. I suggest that individual pieces of evidence, be they experimental results or parsimony claims, have been required to do too much work. I also propose ways of thinking about them that allows each individual piece to play a smaller role.

1.3.1 Defaults and the Hunt for Decisive Evidence

The two features of the debate that were just discussed demonstrate two implicit assumptions that have structured things. These assumptions have proved unproductive. The first assumption is that a parsimony claim, whichever it is that actually applies, establishes a default that must be experimentally *ruled out* before theoretical progress can be made. The second is that experiments that fail to provide *decisive* evidence for one model fail to provide *any evidence at all*. These (usually implicit) assumptions work together and display a common attitude toward the underdetermination problem: we need a single decisive *piece*—an argument or experiment that can break through underdetermination. The effect is to set the stakes for a single consideration too high, asking it to do work that it simply cannot do.

The first assumption, which treats the more parsimonious model as a *default* which must be ruled out before some alternative can be considered, is often employed as a workaround to underdetermination. If the data are indecisive, assume the simpler hypothesis. There are lots of ways that a parsimony claim could be interpreted, but this particular interpretation, which I call 'the default framing' (Dacey 2023) is often motivated by appeal to the procedures of null hypothesis significance testing (NHST). NHST is the dominant statistical method in the field, and it works roughly as follows. An experiment is designed to test between two hypotheses, one of which is named the 'null' hypothesis (usually "no effect"), and the other the 'alternative' hypothesis (usually "some effect"). The null is tested against the alternative. The null is not rejected unless there is, statistically, less than a 5 percent chance the result would be observed if the null were true. If this is found, the null can be rejected. If not, it is not. On a flat-footed reading (which would be contentious, but I'd argue is common in practice), the null is assumed until it can be ruled out, at which point the null is rejected and the alternative accepted.[8] The 5 percent standard is intentionally set low to avoid false positives over false negatives. There are a number of reasons why an experiment might fail to produce a result; there are many variables that are irrelevant to the actual hypotheses under test, that experimenters might fail to control. As a result, the null hypothesis enjoys a kind of privilege over the alternative (Godfrey-Smith 1994).

It's possible to think of model evaluation as a straightforward extension of this process, where parsimony claims such as Morgan's Canon establish a 'null' that must be ruled out (e.g., Andrews and Huss 2014; Bausman and Halina 2018). In fact, NHST and Morgan's Canon align neatly in cases of success testing, in which an experimental result would demonstrate successful performance of a task (as in all of the experiments described above). In these cases, a null result of "no effect" would be the result that is usually taken to be predicted by a hypothesis positing a simpler psychological process.[9] This is because "no effect" is a failure to accomplish the task, and the simpler process is supposed to be the one that is incapable of the task. The more complex process, such as mind-reading, would make the animal capable of performing the task. So, if "no effect" is granted the privileged status of null, it might seem natural for the hypothesis positing the simpler process to receive it as well. This is exactly what Morgan's Canon suggests. Thus, we could take Morgan's Canon to be directly implied by NHST, or even as simply a redescription of NHST, in such an experiment. This makes for a particularly rigid, broadly applicable version of Morgan's Canon because the rules of statistical inference are not supposed to be flexible. Below, I will argue that this reading of Morgan's Canon (and any other parsimony claim so interpreted) inappropriately conflates statistical evaluation with theory choice.

The same general worry applies even when parsimony claims do not align as neatly with NHST as Morgan's Canon often does. For instance, proponents of chimpanzee mind-reading take the more complex process to win on parsimony because it requires fewer processes (process count parsimony). However, even in these cases, the parsimony claim can still establish a default.

---

[8] This reading is perhaps simplistic in ways I will articulate below, but it's not wholly wrong. For example, Popperian objection would claim that neither the null nor the alternative is ever actually accepted. The null can be ruled out or not ruled out, that is all. I don't find this reading all that attractive as a norm for science or accurate as a description; scientists do and should accept hypotheses.

[9] Setting aside, for now, the implication from the logical problem that both hypotheses *could* predict either result; this is the assumption that usually motivates experiments.

28 UNDERDETERMINATION

In this case, the default status comes from the difficulty of interpreting negative results. If a parsimony claim is taken to support the attribution of a more complex process, but the null statistical hypothesis is aligned with the simpler process, then the only 'evidence' one could get *against* the parsimony claim is "no effect." However, it is commonly thought that negative results provide no evidence at all (or, at most, very weak evidence). This fits with a Popperian reading of NHST, according to which one never accepts either hypothesis but can rule out the null if there is a statistical effect discovered. There are also pragmatic reasons why negative results are especially hard to interpret in comparative psychology, which we have encountered: animals may not be motivated to perform a behavior in a laboratory setting, they may interpret the task differently than intended, and so forth. Even so, if we start from a parsimony claim that sets the more complex process as a default, and then discount negative results as sources of evidence, we set ourselves up to never find empirical reason to overturn the parsimony claim and accept the hypothesis positing the simpler process.

This, then, is the first assumption: that parsimony claims establish a default which experimental evidence must override. This applies on both sides of the mind-reading debate. The reasoning is different on each side, but in both cases it can be traced to NHST and the way it structures statistical inference.

The second assumption is that a finding must provide decisive evidence if it is to provide any evidence at all. The emphasis in the literature on attempting to describe an experiment that would avoid the logical problem is based, at least in part, on the thinking that only the right sort of decisive experiment, which attains a positive result, could provide direct empirical evidence about an animal's mental capacities. This is evident in the overall trajectory of the debate, as experiments such as those mentioned above are entirely discounted once it is decided that they don't rule out behavior-reading explanations. It is also sometimes suggested in the way authors discuss evidence. For example, Penn and Povinelli say:

> in order to produce experimental evidence for [theory of mind] one must first falsify the null hypothesis that the agents in question are simply using their normal, first-person cognitive state variables as defined by equation. (2007a, 734)

They go on to assert that the Hare et al. competitive feeding paradigm cannot "even in principle" produce evidence for mind-reading because of the logical problem. Along the same lines, Heyes says:

I argue that there is still no convincing evidence of mental state attribution in non-human animals, and that most current methods of investigation do not have the potential to provide such evidence. (1993, 177)

She later clarifies that she means "unambiguous evidence" (184) or results that could be attributed to a mind-reading capacity but not to behavior-reading. The implication is that evidence needs to be "unambiguous," "convincing," "decisive," and to "rule out the null hypothesis" in order to be evidence at all.

These authors are generally clear that they don't expect any experiment to *prove* one side or the other is correct, or to provide a *crucial experiment*. However, I argue that, in practice, this approach to experimental evidence comes too close to such a demand. The evidential burden that individual experiments are asked to carry is too high.

So individual parsimony claims and individual experimental results are each being asked to carry too much weight. These two assumptions are mutually supporting. Whether the requirements of a single experiment are too high because of high standards set by parsimony claims, or parsimony claims end up establishing defaults because the requirements for a single experiment are too high, the picture is the same in the end. Likely there is something of a feedback loop, and the fact that they fit together so well is a reason the overall approach has been so common.

This overburdening of evidence is the key issue that I address in this chapter. A major part of the problem is that there is a lack of systematic alternatives on offer. I hope to sketch one here (and over the course of the book). We can take the term "evidence" to refer to anything that provides rational reason to increase or decrease one's credence in a particular claim. So, in principle, there can be evidence that is not 'decisive,' 'convincing,' or 'unambiguous.' But the problem remains of articulating, more concretely, what this means. I begin this project by explaining how it is that the evidential expectations for each of parsimony claims and experiments can be reduced. I call this 'modesty' about each.

1.3.2 Modesty about Experimental Results: Separating Statistical and Substantive Hypotheses

As Bausman and Halina (2018) argue, it is a mistake to treat models of cognitive processes as a kind of statistical null (they call this a *pseudo-null*). Actual statistical nulls concern the effects of variables in the specific context of an experiment, while pseudo-nulls are evaluated outside this context. In contexts

30 UNDERDETERMINATION

like the chimpanzee mind-reading debate, statistical nulls in an experiment are chosen for pragmatic reasons (having to do with the difficulty of controlling all possible variables of influence) that only superficially and accidentally map onto the preference for simpler psychological processes encoded in Morgan's Canon. This framing, they argue, is responsible for the unproductive discussion just mentioned and is based on an unjustified mistake: these models should not be treated as 'nulls.'

To expand on this insight and its implications, I have described the problem using the distinction in statistics between *statistical hypotheses* and *substantive hypotheses* (Dacey 2022a, 2023).[10] This language better highlights the source of the mistake and, thus, how it can be addressed. Statistical hypotheses are hypotheses about the distribution of some trait across a population. For example, a statistical hypothesis in the Povinelli et al. (1990) knower/guesser experiments might be that chimpanzees are more likely to search for food indicated by the person who hid the food than they are a person who was absent at the time. A substantive hypothesis makes claims about the causal structure of the processes responsible; about the cognitive capacities that are driving the behavior of the chimpanzees. For example, that chimpanzees attribute states of seeing as well as states of knowing, and they understand that others who *see* certain things will later *know* about them.

The default framing, and the mistake of requiring too much from experimental results, comes from a failure to appropriately separate these. An experiment directly tests statistical hypotheses; it does not directly test the substantive hypothesis. The implication that success or failure of the statistical hypothesis has for the attending substantive hypotheses is a second, separate claim.

The Povinelli et al. knower/guesser experiments are an object lesson about how this step from statistical result to evaluation of a substantive hypothesis can be complicated. The statistical hypothesis that chimpanzees begged more often to the human who hid food was intended as confirmation of the substantive hypothesis that they attribute states of seeing and knowing. However, on more careful evaluation, the statistical hypothesis remains true, but that substantive hypothesis may or may not be the right explanation. It might be that the chimpanzees simply learned to choose the person who hid the food, without any actual understanding of their knowledge states. This is an especially clear case, but in general this slippage between statistical and substantive hypotheses is always possible because of underdetermination. Substantive

---

[10] I have repeatedly spoken of "hypotheses" without either qualification. In general, when I do so, it can be read as "substantive hypothesis."

hypotheses in comparative psychology tend to be characterized in verbal terms that do not force any specific statistical hypothesis. In other words, it may be that chimpanzees do mind-read, but just not in this way, in this task or context (plug in the species and capacity relevant to your favorite debate here). This is part of the problem (e.g., Yarkoni 2022), but not all of it.

This statistical/substantive distinction reframes underdetermination in a way that helps to address it. Specifically, the evaluation of a statistical hypothesis and a substantive hypothesis are two very different things. Statistical hypotheses are evaluated by the tools of statistical inference. Substantive hypotheses are evaluated subsequently to, and separately from, statistical hypothesis (though, crucially, not *independently* of them). In the evaluation of substantive hypotheses, experimental results can be taken to favor one hypothesis or the other. In addressing substantive hypotheses, they can provide evidence even when we don't take them to be individually decisive (even if we do take the statistical result to be decisive). The solution to underdetermination lies in a holistic inference to the best explanation, so it need not be solved by an individual claim or experiment.

I have characterized this process as a kind of two-stage evidentialism, as depicted in Figure 1.4 (Dacey 2023). Fitzpatrick (2008) proposed evidentialism as an alternative to overly strong forms of Morgan's Canon (such as those that create a 'default,' as described above). The core insight for now is that researchers should withhold judgment until there is sufficient evidence. Generalizing this, one can only evaluate a substantive model once all of the evidence is at hand, and one should only express a degree of confidence in that

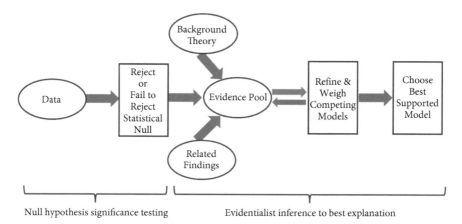

Fig. 1.4 Schematic for two-stage evidentialism, from Dacey 2023.

32 UNDERDETERMINATION

conclusion that reflects the balance of the evidence. That is, one might express a very weak preference for one substantive hypothesis or a relatively strong one, or withhold judgment entirely. I'll connect this to the mind-reading debate more specifically below, but this big-picture framing will be essential to understand my claims there. I move on now to address the overweighting of parsimony claims.

1.3.3 Modesty about Parsimony Claims: Empirical Parsimony

Authors are often led to overweight parsimony claims in this context because they take them to be extra-empirical—that is, outside the reach of empirical test. This understanding of parsimony often arises when it is taken to be a property of the formal, or mathematical, structure of a model. For example, parsimony might be measured by counting the free parameters in the mathematical expression of a model. There are places for this *formal* conception of parsimony in science. However, in this context, I have argued that it is unproductive and should be replaced with an *empirical* conception (Dacey 2016a). Empirical parsimony claims are claims that one model is more plausible than another because of background theory—in this case neuroscience and evolutionary biology (Sober 1990). The parsimony claims operating in this debate are best interpreted as claims of empirical parsimony.

There is a long tradition in the philosophy of science attempting to grapple with the ubiquitous tendency for scientists to prefer simpler theories, and the fact that this tendency *does* seem sensible and useful. A number of these proposals suggest a blanket reduction on the 'things' one posits in the world (e.g., Quine's [1948] famous preference for desert landscapes), or a blanket reduction of terms in an equation. Views like these cannot (or don't even try to) explain how it is that parsimony principles can make us more likely to accept true theories. I don't take these views of parsimony to be helpful in this context. So, instead, I adopt what Sober (2015) calls reductionism about parsimony: parsimony claims can provide good reasons to prefer one theory or another, but they are not rock-bottom principles. Their justification lies either in epistemic considerations (that is, reasons to believe one theory is more likely to be true) or pragmatic considerations (that is, reasons to believe one theory is more useful or manageable).

Empirical parsimony claims, such as those employed in comparative psychology, have the best claim to epistemic justification. Empirical parsimony claims single out a specific feature of the target system that a model is more parsimonious about. Because different claims target different parts of the system, there are a number of different kinds of claim that can arise (Dacey 2016a). Both of the parsimony claims discussed above are like this. Morgan's Canon is the most common, and most widely discussed, dictating that authors prefer models that posit processes that are themselves simpler. The other claim, which prefers hypotheses that posit a smaller count of processes in the mind overall, I've called *process count parsimony*. This often clashes with Morgan's Canon because it can be the case that one more complex process can do the same work as several simpler processes. These two varieties are perhaps the most important to the mind-reading debate, but several others have also played a role. For example, parsimony of processing demands dictates that we prefer hypotheses that require less energy. This will usually align with Morgan's Canon because it's likely that simpler processes (on the functional level) will be less expensive (on the neural level), but it is worth keeping separate. Finally, parsimony of performance-mechanism mapping suggests that we should posit similar mechanisms to explain similar phenomena across species, especially if those species are closely related (Sober 2012; more on this in Chapter 4). These claims are summarized in Table 1.1.

The fact that these parsimony claims pick out different features of the target system indicates that they are claims of empirical parsimony; they are claims about the systems themselves, and which type of system we should think is more likely. They are not claims about the formal structure of the models. Recognizing this, and appreciating the way that empirical parsimony claims can be justified, give them a more productive role in the debate.

As described above, these parsimony claims in the mind-reading debate simply clash. Each side claims parsimony for itself, and dismisses the other

Table 1.1 Varieties of parsimony that have been discussed through the chapter (see Dacey 2016a for more).

Formal parsimony: Prefer simpler formal/mathematical models.
- Prefer model with fewest variables or free parameters.
  Empirical parsimony: Claim that a model which limits a certain kind of posit is better supported by background theory.
- **Morgan's Canon:** Prefer hypothesis positing simplest psychological processes.
- **Process count parsimony:** Prefer hypothesis positing smallest number of psychological processes.
- **Parsimony of performance/mechanism mapping:** Where behavior is similar, posit the same explanation.
- **Parsimony of processing demands:** Prefer hypothesis requiring least energetic or processing demands.

---

34 UNDERDETERMINATION

side's claims, but there is no way to engage or gain traction in discussion about what actually counts as parsimonious and why that matters. This makes sense if each is viewed as a formal constraint with some extra-empirical justification. Which parsimony claims apply, and which are more important, are *a priori* facts on which the evidence has no bearing. If authors on each side take their parsimony claim to establish a default, or a burden of proof, they are not even starting from the same place in evaluating the empirical evidence.

In contrast, claims of empirical parsimony are empirical claims based on empirical evidence. They can be argued for and against, and can be granted whatever weight the empirical evidence behind them supports. Understood as such, they are placed on the common ground of evaluation based on background theory, usually evolutionary biology and neuroscience. More concretely, this means two things. Firstly, individual parsimony claims will be pretty weak because the empirical arguments themselves are probabilistic and often quite tenuous. Secondly, each must be justified by connecting the current discussion to background theory.

Wrapping up this section, I have articulated understandings of experimental evidence and parsimony claims which grant them much more modest evidential weight. Each of experimental results and empirical parsimony claims applies in the second stage of the ordered, holistic inference to the best explanation described at the beginning of the section. Each individual experimental result and parsimony claim can provide only a very small piece of evidence for one side or the other. As a first attempt, we can think of each piece as a small weight added to one side or the other of a scale. Whichever side has the balance of the evidence is the better supported hypothesis. So, for the big question of chimpanzee mind-reading, each side is one of the 'umbrella hypotheses' of mind-reading and behavior-reading. However, I think we can get a much better sense of how we might progress if we put more structure on things.

1.4 Models and Scaffolds

In Section 1.4 I expand on my view of evidence in the debate and suggest new ways that models can scaffold our evaluation of evidence and understanding of cognitive processes. Evidence can contribute to our understanding of what is happening in a process in ways that do not directly contribute to one side or the other of that big question. However, to do so, we need ways to scaffold evidence. This can open new ways to make progress in understanding chimpanzee social cognition.

1.4.1 Narrowing Options

Andrews (2014) describes progress in comparative psychology as a kind of calibration, which has operated by narrowing the range of viable theories. The brief literature surveyed above displays exactly how this has worked in the chimpanzee mind-reading debate as the space of live options has narrowed. The significance of this progress has arguably gone underappreciated in the debate.

Premack and Woodruff (1978) originally asked whether chimpanzees have a *theory* of mind. Theories are generally seen as substantial things, requiring the ability to reason about mental states across contexts and tasks. It is a *theory* because it involves positing and reasoning about unobservable entities (namely, mental states like belief and desire).[11] The field has narrowed its focus since then. Woodruff and Premack (1979) themselves began this process by narrowing their focus to representations of others possessing perceptual states like 'seeing' rather than a general theory of mind. Later, after the competitive feeding experiments (Hare et al. 2001), much of the focus has fallen on competitive rather than cooperative contexts. If we take this narrowing literally, the scope of the supposed capacity is substantially narrowed. More recent reviews often divide mind-reading attributions between attribution of perceptual states, goals, and beliefs (e.g., Lewis and Krupenye 2021; Tomasello and Moll 2013). Perceptual states and goals have been the typical targets of research, especially in competitive settings, though perhaps this is shifting after the Krupenye et al. (2016) false belief experiments.

Theoretical developments have also narrowed options. Povinelli has argued most emphatically against a chimpanzee theory of mind, where 'theory' is read in the rich sense. That is to say, we can take a theory to include a set of unobservable entities that interact with one another, and only have indirect influence in the world. These theoretical entities are crucially interdefined: an understanding of the theoretical entity of gravity requires an understanding of the theoretical entity of mass, and so forth (e.g., Povinelli 2020). In any such case, there is always a thinner understanding possible: one might know that unsupported objects fall without understanding this system of unobservable entities. If so, one could make some gravity-related predictions, but would not have a theory of gravity. We could say the same thing about mental states: that

---

[11] They contrast this with two alternatives—a simpler 'associationist' mechanism and an empathy-driven perspective-taking approach. Note that this latter option would likely count as mind-reading by current standards.

they are unobservable theoretical entities that only make sense as part of such a system of unobservable entities. If so, then anyone with the relevant theory would need to be able to apply it with reasonable generality, and would need to be able to apply most of the other related concepts. Otherwise, we might be able to make some mind-related predictions without having a theory of mind. I take Povinelli's arguments, which are based on systematic failures of chimpanzees to accomplish various tasks, along with the application of Morgan's Canon, to show that it is more likely that chimpanzees do not possess such a theory of mind. Nonetheless, it may be possible that chimpanzees represent others as possessing certain dispositional states that would reasonably count as mental state attributions. This may fall short of a full-fledged *theory* of mind, while still counting as a kind of mind-reading.

On the other end, very simple forms of the behavior-reading hypothesis have also been ruled out. This includes classical conditioning, simple sets of rules directly linking observed movements by conspecifics with appropriate response movements, or anything that doesn't involve some reasoning ability by the chimpanzees.

In response to these trends, some theorists have worked to develop theoretical options in the 'middle ground' between overly sophisticated theory of mind and overly simplistic behavior-reading.

For example, Heyes (2014) proposes "submentalizing" as a version of a behavior-reading hypothesis. Submentalizing is a suite of domain general capacities that "look like mentalizing" (her word for mind-reading) and "provide substitutes and substrates for mentalizing in everyday life" (140). In response to the Krupenye et al. (2016) false belief variant, she argues that apes submentalize rather than mentalize (Heyes 2017). Her submentalizing explanations focus on known processes and capacities that can aid social behavior. So even though these do not involve representations of minds and may not even be systems that evolved specifically for social contexts, they can do important work in social cognition (and indeed, she thinks they do day-to-day work in humans as well, and explicit mentalizing is very rare). In contrast, Butterfill and Apperly (2013) propose a theory of 'minimal mind-reading' that they take to allow successful performance on a number of mind-reading tasks (including false belief). They don't take it to require the actual representations of propositional attitudes, instead representing more limited (but still mind-like) states in others, such as goals, 'encounterings,' and 'registerings' of objects in the environment.

These proposals were developed in response to findings in the literature. It has often been objected that behavior-reading hypotheses are perniciously ad hoc (e.g., Lurz 2011 Halina 2015). Usually, these criticisms target the 'complementary behavior-reading hypothesis' strategy outlined above, where a behavior-reading hypothesis is generated after an experiment that purported to show mind-reading by simply restating the attribution in non-mentalistic terms. I note this here because I think it is important to be clear that, whatever merits this concern might have, the actual *timing* of hypothesis generation should not matter. Both the mind-reading and behavior-reading hypotheses have been, and should be, revised in light of incoming evidence. These revisions must be weighed against the existing empirical evidence, including plausibility based on background theory, and if they make sense in context, it doesn't matter when they were decided upon (see the "refine and weigh competing models" step in Figure 1.4).

This process of revision and refinement is ongoing. Unfortunately, things will only get murkier as we progress from the current state of play. Note that the options not only narrow but also get closer to the boundary line of what we might count as a mind-reading capacity. I take this to be intentional. Butterfill and Apperly call their view "minimal" mind-reading because it is supposed to be the bare minimum. Alternatively, Heyes refers to her view as "*sub*mentalizing", and signals that it does bear some similarity to real mentalizing (it could be called by any other name, for instance, it is not simply "non-mentalizing"). As live options become more similar, it raises two problems. First, it makes it harder to tell exactly what the key differences are between candidate hypotheses. Which should count as mind-reading hypotheses and which not, and why? Second, it makes it more difficult for any single experiment to tell them apart. This, arguably, exacerbates underdetermination.

The solution, then, is to think more expansively about the kinds of progress we can make without answering the big question. Andrews (2014) is right that calibration is one type of progress. So how else can we make progress?

1.4.2 Cognitive Cartography and Refining Models

The progress of refining available models can be furthered if we continue to recognize it as a *central* purpose of experiments. Experiments are not merely evidence on one side or the other of the big question. An experimental effect can tell us one thing that a capacity does or doesn't do. As such, each experiment need not 'test' or 'refute' models (let alone umbrella hypotheses such as mind-reading or behavior-reading). Instead, we can take experiments to set constraints that any possible model has to meet: it should predict *that* result under *those* circumstances.

As such, we can continue to make progress understanding the process without settling the big-picture questions. This progress is made not merely along a single scale of perceived 'sophistication' and is not merely a narrowing of the space of options. It can be a shift in the space of options (as seen in the move to competitive rather than cooperative contexts in mind-reading research), and the space itself is multidimensional. As such, I suggest thinking of it as a kind of *cognitive cartography*: mapping the underlying cognitive capacity, one experimental effect at a time.[12]

I'll discuss two general frameworks for thinking about this kind of project. In both cases, the goal is to advance the progress described in the last section: refining models to better match existing evidence and be more responsive as new evidence comes in. As we learn more about the 'shape' of this capacity, its strengths and weaknesses, the contexts in which it succeeds and fails, then we can make some decision about whether it ought to count as a mind-reading capacity. The only way to actually discover the overall shape of the capacity is with a large collection of experiments across contexts and task variants. To narrow in on one type of experiment, even if one doesn't intend that as a single crucial experiment, is to miss this key point. It's not about a particular experiment, or even a particular kind of experiment; it's a systematic program of discovering what the capacity can do in many relevant contexts.

One way to organize this iterative process takes the 'shape' analogy quite literally. Recently, several authors have proposed multidimensional frameworks for characterizing a psychological capacity. I proposed a two-dimensional framework for distinguishing cognitive processes from non-cognitive processes (which would usually be called associative processes, but see Chapter 3). Birch, Schnell, and Clayton (2020) propose a multidimensional framework for characterizing animal sentience, and Starzak and Gray (2021) proposed a three-dimensional framework for understanding animal causal cognition. Models in comparative psychology tend to be differentiated by degrees of flexibility or sophistication. The dimensions strategy is helpful because it breaks the big question into smaller questions, one for each dimension.

In their dimensional model, Starzak and Gray propose the following dimensions: 1) sources of information, 2) integration, 3) explicitness. This same basic

---

[12] This term echoes Hume's (1748/1974) "mental geography." While there are lots of differences in detail, they share a somewhat similar spirit as mostly descriptive projects that come at the early stages of attempting to understand the mind (e.g., Jacovides [2024]).

framework could plausibly be adapted to mind-reading. In the mind-reading context, *sources of information* might include bodily posture, movement, head orientation, and eyes. It may also include geometry of the environment, including things like opacity and translucency of barriers. It could include an ability to project from one's own experience of having been in the same position as the others (see Lurz's suggested experiment, 2011). It may also include an understanding of the conspecific's context or past: how they have tended to behave, what preferences they might have, whether they are likely hungry and so forth. A species scores higher on this dimension when it can gather information from more sources. *Integration* would measure an ability to integrate these different sources of information into a coherent representation or understanding of the situation. Transfer tests would be the best measure here (e.g., Heyes 1993), but we might also look at tests that provide cues of different kinds and see if chimpanzees are able to use each kind in solving a single task. Finally, *explicitness*, as Starzak and Gray characterize it, is the ability to use the resulting information in different ways. For example, the ability to use complex social information for competitive but not cooperative tasks shows less explicitness than if both were possible. Other tasks that have been suggested include mere predictions of behavior (as in the Krupenye et al [2016] false belief variant), and attempts by chimpanzees to explain unexpected behavior they observe (suggested by Andrews 2012). A tendency to seek out information from a certain missing source would also indicate strong integration as well as explicitness in the form of an ability to explicitly assess confidence in one's attributions based on existing evidence (see Buckner's [2014] suggested experiment).

This particular three-dimensional model may not be the best framework for understanding mind-reading. The choice of model is pragmatic as much as it is principled. Each of these three dimensions is quite abstract and so might not break up the questions at hand into small enough problems to make progress. If so, introducing more dimensions might be better, or relativizing the 3-D space to a specific version of the task—for instance, one for competitive contexts and another for cooperative. It is an open question which exact version will be most helpful, and that is what should decide the matter.

The second cartographic strategy maps the overall response profile of the capacity and uses this information to infer the contents of the representations involved. The idea here is that, while a particular experimental result might be predicable by either a mind-reading or behavior-reading hypothesis, the overall pattern of responses may be more plausibly explained by one or the other. I interpret this strategy in light of Buckner's (2014) insight that the logical problem is at its heart a problem of *psychosemantics*: a problem of determining the content of the representations that chimpanzees use as they engage with one another. The question, in this framing, is about the representational content of the states in the mind of the chimpanzee.

Models of mind-reading capacities often carry systems-level claims, and so discussions of whether chimpanzees are able to mind-read are often framed around the question of whether a dedicated mind-reading system is present. Butterfill and Apperly (2013) describe minimal mind-reading as a 'type one' automatic system in contrast to a 'type two' controlled process, while Heyes (2014) takes the mind-reading hypothesis to require a specific dedicated 'module' or system. Penn and Povinelli's influential characterization of the mind-reading ability depends on a sophisticated architecture where mind-reading is performed in a new, emergent system (2007a). Systems-level claims like this can be helpful because the hypothesis that mind-reading is subserved by a particular system might open new ways of looking for that *system*: for example, if we build mind-reading into a dual systems hypothesis, as do Butterfill and Apperly, we might be able to look for signatures of automatic versus controlled processing (e.g., fast vs. slow, effortless vs. effortful), which bear on the model without requiring direct evidence about what the purported 'mind-reading' system is capable of. However, these inferences depend on the specifics of one's preferred system architecture model, and most of these models are speculative. Moreover, if we see mind-reading as a psychosemantics problem, the core question remains regardless of the number of systems present: is the system actually representing mental states? For example, we might have good reason to think that there is a dedicated 'social reasoning' system in chimpanzees, which is homologous to our own mind-reading system (see Chapter 4). But that system still might or might not represent mental states. To address this question, we must gain a fuller map of the way the purported mental state representation behaves, and what it is capable of doing.

To help get this project started, I will point out some similarities between mind-reading and behavior-reading hypotheses. Both hypotheses take the animals to use perceptual information to make predictions about behavior. Both sides agree that chimpanzees have sophisticated abilities to reason about and predict what others might do in a given situation. As a result, both hypotheses must take chimpanzees to attribute dispositions to conspecifics on the basis of observable evidence, and then act on the basis of the predictions generated. This requires that some more basic perceptual information is processed and passed along to the stage at which the key inferences are made (at a minimum, for example, shape recognition has to flow into face recognition, which then can influence behavior on either hypothesis).

So, in fact, the basic inferential structure is the shared between the two umbrella hypotheses (Figure 1.5). The core difference is whether the disposition chimpanzees attribute to one another amounts to a mental state attribution or not. A chimpanzee might expect certain responses because they attribute a mental state (see Figure 1.5a). Alternatively, they might attribute behavioral dispositions to various self-moving entities without taking them to be minded, as we often do (e.g., machines, single-celled organisms, perhaps plants and some simpler animals; see Figure 1.5b). From the perspective of a comparative psychologist, the issue is whether the dispositions that chimpanzees attribute to one another are flexible and sophisticated in the ways we expect from mental state attribution (see Figure 1.5c). There might be particular signatures; false belief attribution is a strong candidate, for example, and attribution of some sort of teleology (actions directed at ends or goals) seems requisite. But in the actual observed behaviors, these will emerge in the overall pattern of responses across task variants and contexts.

The structural similarity of the models in Figure 1.5 is, in part, a restatement of the difficulty in telling the two umbrella hypotheses apart. The actual patterns of inputs, inferences, and outputs will differ between models, and getting precise about these predictions is exactly the business of refining the models. However, there might be any number of inference patterns that might reasonably count as either mind-reading or behavior-reading. So, the failure of any

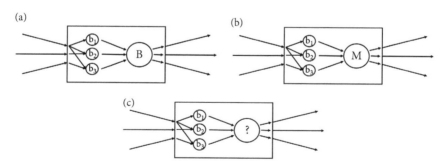

Fig. 1.5 Reframing the difference between mind-reading and behavior-reading. The lower-case b's are representations of simple features of the environment, such as body orientation, geometry of the space, and so forth. The upper case 'B' in 1.5a is an attribution of a behavioral disposition to another agent. 'M' in 1.5b is an attribution of a mental state. Fig. 1.5c demonstrates that the inferential structure can be basically the same between these.

one such model cannot undermine the entire category. As a result, the umbrella categories cannot be decisively tested with any single experiment: we would need to look across experiments. At any given time, one model matching one of the umbrella hypotheses may provide the best explanation for the collective evidence. But the solution will only come with a systematic understanding of the causal and inferential structure of that representation.[13] This requires gathering evidence from many tasks and contexts. Moreover, as depicted in Figures 1.5a and 1.5b, the two models look the same, but the detailed structure will differ based on possible models and as new evidence comes in. Indeed, these could ultimately look quite different from one another. The question, though, is how we identify those differences in the inputs and outputs.

The approach avoids the trap described above, whereby any result that is not decisive evidence is no evidence at all. If we incorporate a specific piece of evidence into the overall network, we are learning from it, whether or not it provides direct evidence for either umbrella hypothesis. However, it might also provide such evidence indirectly; incorporating that evidence into the network might force it into a shape that looks more like one or the other umbrella capacity.

The experimental program this suggests is one in which tasks are systematically varied and responses recorded. This, perhaps, suggests moving away from the 'success-testing' framework most of the literature has used, and simply testing any variant that might provide information.[14] As this goes along, we can compare it to patterns we might consider to be reasonable examples of mind-reading versus behavior-reading. This is a process of continual, iterative refinement and re-evaluation of competing models.

To answer the big question, the field needs to make decisions about exactly what the mind-reading capacity entails. Where exactly is that boundary? What exactly are the crucial differences between a minimal mind-reading capacity and a sophisticated behavior-reading capacity? What kinds of mental states do we think chimpanzees might, or might not, attribute (Boyle 2019)? We cannot expect chimpanzee mind-reading, if it exists, to look all that much like our own. Having more specific models helps, but even then the question can reassert itself. For example, in assessing minimal mind-reading (Butterfill and

---

[13] Buckner argued for a particular Dretskian view of representational content. This might make it look like I am relying on inferential role semantics. I don't think it matters for present purposes. When it comes to gathering evidence for representation, it seems to me that patterns of success and failure in social tasks are a good place to look regardless of one's view of content (the specific experiments one suggests might differ).
[14] As in Taylor et al.'s (2022) "signature-testing" suggestion, though I am wary the 'signature' language suggests a clear individual identifier.

Apperly 2013), it's worth questioning exactly what representations and inferences would count as encountering and registering, and why these count as mental state attributions. We might also question whether the capacities Heyes (2014) describes as submentalizing can combine in ways that actually allow a mental state attribution.

So there remains considerable work to be done. But, crucially, this work can be done without deciding the big question. Thus, various experiments and parsimony argument can play into the various projects described in the last two sections. They need not be decisive to be valuable; thus, we break out of the all-or-none reading of evidential value.

The last few sections have set up a big-picture way of looking at the debate that I think will help make progress. One might reasonably wonder where the chimpanzee mind-reading debate actually stands in light of this framework.

1.5 Summing Up and Stepping Back

I take the big question to remain unsettled. Lurz, Kanet, and Krachun (2014) argue for an "optimistic agnosticism" on the question of chimpanzee mind-reading. As things stand, I think this is basically right. Were I pressed on the issue, I would say that I think it's more probable that chimpanzees are capable of mind-reading in a way something like minimal mind-reading in at least some situations. It seems likely to me that chimpanzees can attribute states that are reasonably characterized as goals and perceptual contact; I am less convinced by the evidence of false belief attribution as things stand. Overall, as a result, I take the mind-reading (umbrella) hypothesis to have the edge. However, my confidence in all these assessments is low, and as such I think the appropriate official position to take is still agnosticism. My goal in saying this is not to muddy the waters or to push ecumenicism for the sake of ecumenicism; I simply intend to model the kind of reasoning process that I see as appropriate in difficult theoretical questions like this.[15]

The question of chimpanzee mind-reading is not likely to be settled by an experimental breakthrough that solves the logical problem. Indeed, even if we address the logical problem specifically, more mundane versions of underdetermination hold. Instead, we should expect continued slow progress coming with new experimental findings and refinement of existing theoretical

---

[15] Note that the decision of when to make a claim is a key place in which values can play a role in science (Douglas 2000), so one might reasonably disagree there as well.

options. I suggest a shift away from taking experiments to be direct tests of substantive models towards a strategy of cognitive cartography which compiles the capabilities and limitations of the dispositional representations chimpanzees form, and a continued discussion of whether those themselves constitute a mental state attribution. This fits into a broader ordered holistic evidentialist strategy in place of the currently dominant default framing.

1.6 Conclusion: Beyond Mind-reading

I focused on the chimpanzee mind-reading debate here because it provides a particularly salient example of underdetermination. However, I take my conclusions to generalize. Firstly, as Povinelli (2020) points out, the logical problem is not restricted to the mind-reading debate. The same problem appears in any debate over an ability to posit a theory made up of unobservable entities, including causal structures, weight, and the like. So anywhere a version of the logical problem holds, my arguments straightforwardly apply. Moreover, I don't think that the logical problem requires a fundamentally different approach than underdetermination more generally. The approach that I have outlined here to address underdetermination in the chimpanzee mind-reading debate can be generalized to help address underdetermination across comparative psychology. There are three main take-home points.

First, we should be modest in interpreting the evidential strength of individual experimental findings and parsimony claims. Parsimony claims ought to be interpreted as claims of empirical parsimony, which means they have some value, but it is much weaker than is often assumed. Experimental results should be taken only to directly bear on statistical hypotheses, with implications for substantive hypotheses requiring further interpretation and argumentation.

Second, experimental results ought not to simply be taken to deliver evidence that counts in favor of one or another substantive hypothesis. It is also, and perhaps sometimes primarily, a piece of evidence that helps characterize the capacity across relevant tasks and contexts. In fact, I even take the sort of diagram proposed in Figure 1.5 to generalize (see Chapter 3). In many cases, an animal is perceiving certain features of the environment, using it to construct a representation of the situation as being a certain way, and then acting on the inferences licensed by that representation. The question is what sort of representation that is, and what inferences it licenses.

Third, in order to settle the big questions about whether a species possesses some capacity, we need to make decisions about exactly what overall patterns of success and failure, what inferences and what indicators, we require for a process to be taken to represent information of a certain kind. We should keep in mind that animal versions of a certain capacity will not usually be merely 'diminished' versions of our own, and so they will lack the ability to make some discriminations, but they will likely make others that we do not. There is a judgment call that needs to be made here, and it will take time.

These lessons apply across the field: to debates about causal reasoning, about episodic memory, self-recognition, metacognition, numerical cognition, and so forth. In all of these areas, underdetermination holds, and as a result researchers face the same core issues. This is a very difficult and demanding process. However, I do not think that is the fault of the framework here; it is the fault of the subject matter. To the extent that this process requires judgment calls, or assessments of large swaths of literature that are likely debatable, these are things that need to happen anyway. This framework makes them more explicit and takes them out in the open.

The inferential gap between observation and theory remains; the goal here has simply been to sketch a plan for a bridge. I said above that underdetermination sets the stage for the other challenges discussed in the coming chapters. This remains true; the six other challenges all arise as we attempt to cross the gap.

2

Anthropomorphic Bias

2.1 Introduction

In spring 2014, the City of Baltimore installed the Inner Harbor Water Wheel to reduce trash pollution in its harbor. It is, in essence, a barge at the end of an inlet to the Inner Harbor, which sifts trash off the top of the water as it flows by (partially powered by a waterwheel). After a video of the wheel went viral, the operators sensed an opportunity: They placed giant googly eyes on the roof of the barge, renamed it Mr. Trash Wheel, and gave it a Twitter account. A social media star was born. Riding on Mr. Trash Wheel's fame, the program has flourished, and there is now a "family" of four Trash Wheels in Baltimore, with two more on the way to operate in California and Panama. Each has a different name, look, and Twitter (or X) personality (and, of course, merch). Mr. Trash Wheel has inspired local beer names, a festival in the Wheels' honor, international coverage, and even a tattoo on at least one fan. Giving a trash barge some cute googly eyes changed the way people engage entirely. It made a meaningful difference for the project of cleaning up the Inner Harbor and perhaps, eventually, cleaning plastic out of the oceans worldwide. This is the power of anthropomorphism.

Anthropomorphism, as a first pass, is the tendency to interpret nonhumans (for our purposes, animals) as if they were humans. Of course, Mr. Trash Wheel himself is all fun and games. Nobody really thinks Mr. Trash Wheel is writing those tweets, and the whole thing is very much tongue-in-cheek.[1] However, if you look up before-and-after photos of the added googly eyes, you will see just how powerfully he *looks* different with those eyes. If there were an actual question about whether he had thoughts and a personality, the eyes might very well tip the scale. In fact, the worry that anthropomorphism might do so unduly, and cause error, has shaped the methods and structure of comparative psychology from its earliest days. But what we make of this worry depends on how we see anthropomorphism.

---

[1] It falls on Professor Trash Wheel to tweet serious facts about environmental science; for more see https://www.mrtrashwheel.com/.

Seven Challenges for the Science of Animal Minds. Mike Dacey, Oxford University Press. © Mike Dacey 2025.
DOI: 10.1093/9780198928102.003.0003

The idea that we see humanness all around us has strong intuitive appeal, and is not at all new. It was put characteristically well by Hume:

> There is an universal tendency among mankind to conceive all beings like themselves . . . We find human faces in the moon, armies in the clouds; and by a natural propensity, if not corrected by experience and reflection, ascribe malice or good-will to every thing, that hurts or pleases us. (1757/2007, Sect. III)

In the modern world we might treat our pets as children, take the traffic jam personally, and think our glitchy computer has it out for us. The idea, then, is that we have a *cognitive bias* to anthropomorphism: an automatic, unconscious, unintentional tendency to interpret nonhuman 'agents' as we would human beings (Dacey 2017a). I call this *implicit anthropomorphism* to distinguish it from other kinds and conceptions of anthropomorphism in the literature. Too often, in literature on animal minds, anthropomorphism is treated as merely a tendency to attribute one of some set of 'distinctively human' traits to animals that do not have them: minds themselves, or consciousness, or perhaps intentions, beliefs, desires, and emotions, or elaborate psychological capacities like mind-reading. The implication is that anthropomorphism has the effect of uniformly inflating the perceived intelligence of animals in comparative psychology. This understanding, as we will see, has dominated attempts to control the influence of anthropomorphism in the science.

However, once we recognize that the concern here is one of cognitive bias in the (human) researchers, the deficiencies of such an approach should be immediately clear. We cannot simply intuit the influence that any given cognitive bias will have; the history of the psychology of bias is one of constant and ongoing discovery of surprising effects. We should expect the same of implicit anthropomorphic bias. There are good reasons to think implicit anthropomorphism has a number of effects in different situations. Really, the way forward is to study, empirically, human tendencies to anthropomorphize. Thus, as we discussed chimpanzee mind-reading in the last chapter, this chapter will discuss mind-perception in human beings.

The psychology of anthropomorphism has been surprisingly neglected, only receiving real interest in roughly the last fifteen years. There are a number of distinct reasons why it might interest psychologists. Perceiving others as 'minded' is a crucial part of moral psychology (Gray, Young, and Waytz 2012;

Waytz, Epley, and Caccioppo 2010). Anthropomorphism has also long been posited as a contributor to religious intuitions and the structure of religious stories and explanations (Shaman, Saide, and Richert 2018). Social robotics has more recently emerged as a research focus (Sheridan 2020). Work in this area aims to improve cooperation and interaction between humans and various machines as they take over aspects of our daily lives, or we find ways to provide robotic companions for the lonely (van Doorn et al. 2017). However, I argue that the study of implicit anthropomorphism, specifically, has been hampered by a lack of the right kinds of implicit measures. Several researchers are trying out possibilities, including a collaboration I am involved in. It is not yet clear if any of these can support a systematic study of implicit anthropomorphism as it might affect comparative psychology. Nevertheless, developing the right methods is the first step.

In this chapter, I first present worries about anthropomorphism as they have manifested in comparative psychology through the years, arguing that the core shared worry is about an unconscious cognitive bias, though this has been underappreciated (Section 2.2). I then argue that this means that existing methods for controlling anthropomorphism in comparative psychology, especially Morgan's Canon, are not up to the task (Section 2.3). After surveying empirical research on the human tendency to anthropomorphize (Section 2.4), I describe my own attempted contributions (Section 2.5). I then step back to look at the place of unconscious processes in scientific reasoning (Section 2.6).

2.2 Worries in Comparative Psychology

Worries about anthropomorphism date to the very founding of comparative psychology as a field. The stage was set in the aftermath of Darwin's work on animal minds. Darwin took the theory of evolution, based on incremental changes over millennia, to indicate that we should expect the mental capabilities of animals and humans to be continuous. While this was a tremendously controversial claim at the time (to say the least), it was enthusiastically taken up by Darwin's followers. They pursued a research program that sought behaviors that would be naturally explained in humanlike terms. Around the turn of the twentieth century, authors such as C. Lloyd Morgan and Edward Thorndike objected to this anecdotal, anthropomorphic method. Morgan's own objections at least partly motivated his "Canon," which continues to be deployed as an inference principle intended to avoid anthropomorphic conclusions (more on this shortly).

Concerns about anthropomorphism, or at least concerns about overstating the intelligence of animals, drove the field through the behaviorist era,[2] and into the cognitivist era that followed. Today, positions have arguably softened and diversified, but it remains a subject of frequent debate. These debates are difficult to follow in part because different authors come in with different purposes and preconceptions. As such, the term 'anthropomorphism' can *mean* many different things.

In comparative psychology, charges of anthropomorphism have often been levied by critics against specific theories that attribute sophisticated abilities to animals. This might include beliefs, desires, consciousness, mind-reading, self-awareness, emotions, various specific intentions, or even intentions *simpliciter*. This use of the term implies that anthropomorphism, as a matter of definition, is an error: the positing of humanlike traits that are not actually present (it is thus a 'thick' term in philosophical parlance, including descriptive and normative/factive elements).

If one wishes to defend a theory against such a charge of anthropomorphism, one has two options. The first is to argue that the theory is not actually anthropomorphic. This, in effect, accepts the framing of the term anthropomorphism as an error but suggests that it does not apply. The second approach is to argue that anthropomorphism in this case is appropriate. The claim could be that the trait posited may be 'humanlike,' but it is not uniquely human, because *this* animal has it. This reply accepts that anthropomorphism is about positing humanlike traits, but denies that it is definitionally an error (i.e., it accepts the descriptive portion but denies the normative/factive part of the definition above).

These debates tend to play out as debates about the *theory* under discussion rather than as discussions of anthropomorphism itself. But there has also been discussion on the metalevel issue of anthropomorphism. Some authors, most notably Burghardt, view anthropomorphism as a legitimate method. Burghardt (1991) describes *critical anthropomorphism* as a deliberate, intentional procedure of putting yourself in the animal's position and attempting to see the world as it does, based on what you know about that species. This claim is certainly controversial (see Wynne 2004, 2007; Burghardt 2007). It also represents a third implied view of anthropomorphism: of primary interest here is

---

[2] Admittedly, the behaviorist position is an awkward fit with the term 'anthropomorphism,' as it was a consequence of their view that humans and animals are not that different. They arrive at this position, though, by arguing that humans are *also* driven by mechanisms more commonly attributed to animals. Nevertheless, the worry about a tendency to inflate perceived mental sophistication was central.

50 ANTHROPOMORPHIC BIAS

not the actual posits one makes but the *process* by which one considers possible posits.

As a result of these varied perspectives, the terminology is often murky. I apply the term 'anthropomorphism' to the process rather than the posit. This approach avoids treating any particular list of posits as 'humanlike,' which I take to be misguided for a number of reasons that I won't fully explore here. Taking anthropomorphism to definitionally include an error makes charges of anthropomorphism circular, insofar as they are arguments against a claim about animal minds; it presupposes an answer to the question of animal minds that is supposed to be at issue. These are problems that we don't encounter if we instead view anthropomorphism as a feature of processes of reasoning or judgment. Most generally, anthropomorphism-as-a-process means that one is thinking about animal minds by engaging the same thought processes they use to think about human minds. This allows that there can be lots of kinds of anthropomorphism; that is, lots of processes that might be, in the important sense, anthropomorphic. Burghardt's critical anthropomorphism involves a kind of internal simulation. We might also take anthropomorphism to involve tokening a representation of a person in one's reasoning process (Shaman, Saide, and Richert 2018). But both of these might be very different than the way Mr. Trash Wheel's eyes give him apparent personality, or we feel that a customer service chatbot is being intentionally evasive, or we come to think a cawing crow is angry with us, or we imagine that a butterfly is approaching us out of curiosity.

Attributing anthropomorphism to the process also sets up my more specific focus on implicit anthropomorphism: *when our implicit folk psychology responds to nonhuman animals the same way it responds to humans.*[3] Implicit anthropomorphism is, I argue, the real core challenge of anthropomorphism for comparative psychology. I say this for two main reasons.

Firstly, and most importantly, I take this to be the worry in the background of all discussions of anthropomorphism. It is shared across theoretical orientations, even if it is not often acknowledged as such. The harshest critics of anthropomorphism worry so fiercely about it because they see a need to protect against a basic human tendency. Kennedy is so opposed to anthropomorphism that he advocates for neobehaviorism in large part because the tendency

---

[3] Thus, the anthropocentrism we find here is less pernicious than any version that circumscribes a set of "uniquely human" traits. It is anthropocentric simply in the sense that we are humans with human perspectives. The idea that this use of implicit folk psychology is *anthropo*morphic simply requires the assumption that our folk psychology is most engaged in interactions with humans, or that it evolved to interact with other humans, and then extended to animals (Caporeal and Heyes 1997).

to anthropomorphize is "simply built into us" (1992, 28). Nothing short of a wholesale change in the way we describe the mind can avoid a pernicious impact on the science, he says. On the other side of the theoretical spectrum, Rivas and Burghardt caution:

> Anthropomorphism is like Satan in the bible—it comes in many guises and can catch you unawares! . . . If anthropomorphism is a natural tendency of human beings, scientists are not immune; lurking unseen, it can compromise efforts in many areas. By using critical anthropomorphism and trying to wear the animals' 'shoes,' we can overcome part of our natural bias and obtain a more legitimate understanding of the life of other species. (2002, 15)

So controlled, intentional (critical) anthropomorphism can be helpful precisely *because* it can help address the implicit forms of anthropomorphism (which we still must beware of). Most theorists likely fall between these positions and still think anthropomorphism is a major problem for the field.[4]

Overall, I take it that the position one takes on the potential usefulness of (some form of) anthropomorphism depends substantially on one's attitudes about the balance of risk and benefits of implicit anthropomorphism. Strict opponents of anthropomorphism see its implicit aspects as distorting, with no upside. Proponents may downplay the risks of implicit anthropomorphism (compared to opponents), such that they are outweighed by other benefits. For instance, engaging our implicit folk psychology might foster a kind of creative insight, or it may make possible a level of understanding that we would otherwise lack (Burghardt 2007). Thus, implicit anthropomorphism is at the heart of debates about anthropomorphism in comparative psychology, even if it is not always explicitly recognized.

Secondly, implicit anthropomorphism is the hardest version of the challenge. Implicit folk psychology is not one thing but a suite of processes, some perceptual, some cognitive, some emotional. As such, just as Rivas and Burghardt say, implicit anthropomorphism can show up in many forms in many settings. Often this is likely fine, but like any unconscious process this can and does lead to biases. When they lead us to err, I call it *anthropomorphic bias*. Really, though, there is not just one bias here: there are a constellation of biases that may look very different from one another. They share some common origins and produce effects of potential concern in the same contexts, as they are

---

[4] There are disciplinary differences here as well, with comparative psychologists typically more negative about anthropomorphism than cognitive ethologists; more on this division in Chapter 5.

52 ANTHROPOMORPHIC BIAS

products of various aspects of our implicit folk psychology.[5] Because of this variety, we cannot identify any criterion on which to readily identify instances of bias. Nor can we escape or 'turn off' the processes that produce them. As with any bias, there may be indirect methods of control, but we are largely stuck with what we have. Kennedy may be wrong in his conclusions, but he is right that implicit anthropomorphism is simply built into us.

For these two main reasons, I see implicit anthropomorphism as the real challenge. The practical question, then, is what to do about it. Unfortunately, existing answers to anthropomorphism have been driven by researchers' intuitions about what they think anthropomorphism might be, rather than understanding what this bias is and how it works. Specifically, the assumption is that anthropomorphism is primarily about inflating the perceived intelligence of animals. While it may be clear why I do not think this is the right view, it's worth understanding how it plays out.

2.3 Animal Smiles, Slouching Shoulders, and Why Morgan's Canon Fails

Morgan's Canon is the most commonly proposed and applied control for anthropomorphism in comparative psychology. I've already discussed Morgan's Canon some in the Introduction and Chapter 1. As the Canon has been interpreted more recently, it dictates that authors should default to the candidate hypothesis that posits the simplest possible psychological process, until one has decisive evidence against it. Simplicity, here, is generally conceived in functional terms, having to do with the sophistication or computational complexity of the process itself. This is usually operationalized in terms of flexibility of the behavior that process produces. For example, as discussed in the last chapter, Morgan's Canon would tell against the hypothesis that members of any other species attribute mental states, in favor of behavior-reading hypotheses that do not posit this ability.

So, the thinking goes, when we have two options on the table, one of which posits a more sophisticated cognitive process and one a simpler process, our tendency to anthropomorphize might lead us to systematically err on the side

---

[5] As such, I take it to be worthwhile to continue referring to them under the collective heading 'anthropomorphic bias'; one must be careful, though, not to expect anything like a natural kind. There are likely boundary cases of biases that may not clearly count as either anthropomorphic or not in this sense. I don't think this matters for the core claim here; in any case the goal will be to understand the bias and its implications better.

of over-attribution. This tendency might make *errors of commission* common, where we posit sophisticated abilities that are not present. This is the vision of anthropomorphism whereby its influence is to uniformly inflate the perceived intelligence of animals. If this were true, then a principle that pushes with near-equal force in the opposite direction would be a reasonable corrective. Perhaps Morgan's Canon can do that job; whatever tendency to err in the direction of overestimation might be counteracted by an equally strong preference for hypotheses positing simpler psychological processes (see Figure 2.1).

De Waal (1999) and Sober (2005) frame the evaluation of this conception of Morgan's Canon as a trade-off between errors of commission (just described) and possible *errors of omission* created by Morgan's Canon itself. They refer to the errors of commission as anthropomorphism (thus, using the term in one of the ways I chose not to) and errors of omission as *anthropodenial*. Errors of omission are just as much errors as are errors of commission, they point out. The blanket use of Morgan's Canon doesn't reduce errors but, instead, amounts to a chosen preference for errors of omission. Indeed, Morgan's Canon is used so often and so forcefully that it arguably *increases* the overall error rate. If there is no reason to prefer one of these errors over the other (which I'll consider more in Section 2.6), the goal should be to reduce the overall error rate, and Morgan's Canon is counterproductive to that end.

De Waal and Sober are right that Morgan's Canon is too forceful to play this role and, as a result, has likely produced more errors than it has corrected (at least, in recent years). However, I think the problem is more fundamental than

	Theorists Claim: Absent	Theorists Claim: Present
Absent	✓ *accurate*	✗ Anthropomorphic error
Present	✗ Anthropodenial	✓ *accurate*

Anthropomorphism ▬▬▬▬▶
◀▬▬▬▬ **Morgan's Canon**

Fig. 2.1 Anthropomorphism, Morgan's Canon, and the trade-off of errors. The table depicts the space of errors when deciding whether to claim members of some species have some sophisticated mental capacity. If it were the case that anthropomorphism produces a uniform tendency to overestimate animal intelligence by making errors of commission, and Morgan's Canon produced a nearly equal pressure in the opposite direction (represented in the arrows below the table), then Morgan's Canon could effectively correct anthropomorphic error. However, as I argue in this chapter, neither antecedent of that conditional is true.

54 ANTHROPOMORPHIC BIAS

this. Overestimation of animal intelligence by positing sophisticated processes is likely only *one of* the errors produced by the human tendency to anthropomorphize. Even if we accept that overestimation is the most common type of error, it is not the only one.[6] Oftentimes, our tendency to see animals as like us leads us to misunderstand behaviors in ways that have nothing to do with intelligence or sophistication.

The primate 'smile' is perhaps the clearest example (indeed, this is the prime case for my studies discussed below). When humans see a primate such as a monkey or a chimpanzee grinning, it looks like a big, toothy smile. The monkey *looks* happy. However, this facial expression usually expresses something more like fear or anxiety.[7] We see happiness in the chimpanzee because we intuitively interpret the grin in the same way we would on a human being. Thus, in this case, interpreting the chimpanzee grin in the same way we would interpret a human expression leads us to get it wrong. However, this is not a mistake that has anything to do with the intelligence of the animal; whichever hypothesis one choses, that the animal is experiencing happiness or fear (or if one really insists, proto-happiness or proto-fear), one attributes the same cognitive sophistication in the animal. Many other cues we see in animals might have a similar effect. A baby penguin might slouch its shoulders in a way that makes it look sad. A skate pressed up against aquarium glass can look deliriously happy, with a goofy smile and nostrils appearing like eyes. People also often misinterpret their own pets. Dogs that roll over are making a show of submission, not just playfully seeking a belly-rub because it feels nice (though that doesn't mean they don't enjoy the rub, and we are not helpless identifying their emotions: Schirmer, Seow, and Penney 2013). These are just examples, but they all involve implicit anthropomorphism producing errors that do not impact the perceived intelligence of the animals.[8]

---

[6] I acknowledge that the degree to which one sees my arguments as important depends on how one thinks these break down. For example, if errors of commission are 90 percent of the errors that implicit anthropomorphism produces, then Sober and de Waal have it close enough to right that my arguments here don't matter much practically. However, I see no reason to think errors of omission dominate to this degree, and plenty of reason to think other types of error are quite common.

[7] This grin is likely homologous to the human smile (Waller and Dunbar 2005; Parr and Waller 2006), but the signaling function is different. Note that in chimpanzees and (especially) bonobos, the grin may sometimes signal pleasure but more often signals fear, nervousness, or hesitation (de Waal 1988). Thus, the function of the expression may get more humanlike as we get phylogenetically closer to humans, though even in our closest relatives it is still different.

[8] Individuals will differ in how intuitive these examples seem, especially given informed experience with members of these species; most researchers likely do have better intuitions about their species of study than the general public. However, we should be careful not to assume that experience alone automatically makes for better intuitions.

Implicit anthropomorphism can also lead us to *miss* intelligent actions. If the intelligence behind an action is different enough from our own, we might simply fail to see it. The cognitive abilities of cephalopods, especially octopuses, have recently become a topic of significant focus (e.g., Godfrey-Smith 2016). It is likely, though, that part of the reason we failed to recognize it for as long as we did is that they are so alien in form and lifestyle. Moreover, these can interact; if we misread the emotional state of an animal when performing some action, we might misread the stakes from the animal's perspective and so misunderstand its goals and intentions. Rivas and Burghardt (2002) capture many examples like this under their term "anthropomorphism by omission." They argue that many cases of anthropomorphism underestimate intelligence of animals because they involve the presumption that the suite of capacities an animal has must be a subset of the capacities that humans have. In reality, though, animals have many capacities that we do not—perceptual capacities, behavioral repertoires, and even reasoning abilities.

Often, cognitive biases can be seen as resulting from a particular stereotype in thought. I think the stereotype applied by our intuitive folk psychology, as it feeds into anthropomorphism in a scientific context, is best summarized by the slogan "all intelligence is like human intelligence" rather than "all animals are like human animals." Thus, intelligence that we see will be interpreted as similar to human intelligence, and intelligence that is not like human intelligence may not be seen at all. The first step in truly addressing this is recognizing it as a cognitive bias. One of the challenges of dealing with cognitive biases is that the effects they have can be unpredictable. The only way to understand cognitive bias is to study it empirically.

So, one of the mistakes in the idea that Morgan's Canon can control anthropomorphism is the implied assumption that we can just intuit the impact this cognitive bias has. We cannot, and there is good reason to think the answer presupposed by Morgan's Canon is wrong. Not all instances of anthropomorphism involve inflated perceptions of intelligence. And in the cases that do not (which may well be the majority of cases), Morgan's Canon is completely unrelated to the effect of the bias. Indeed, looking at work on social biases, interventions that demand a certain decision in the face of bias, as Morgan's Canon does, can sometimes *increase* bias (Dacey 2017a).

Another problem with Morgan's Canon is that it operates primarily at the moment of hypothesis evaluation. But cognitive biases pervade our lives. They operate all the time and have the chance to influence thinking at any stage of any reasoning process. This applies here as well. Implicit anthropomorphism can influence the candidate models that we come up with in the first place

56 ANTHROPOMORPHIC BIAS

(more next chapter). It can influence the kinds of tests we devise, since we may have to intuit what task might engage the capacity under question in the right way. It can influence which questions we see as worthy of asking in the first place. It can influence what behaviors we literally see when observing an animal. It can also, of course, influence the way we evaluate candidate hypotheses given a certain set of data.[9] Though Morgan's Canon admittedly indirectly influences other aspects of science, the aim is to generally direct one towards accepting, in the end, the simplest hypothesis possible.

I should say a few words about the evaluation of models, given the picture that I outlined in the last chapter. There, I argued that we should separate the quantitative evaluation of statistical models from the qualitative evaluation of the substantive hypotheses. In particular, I argue that models should be evaluated with a holistic, qualitative, 'evidentialist' inference to the best explanation. One might point out, fairly, that this procedure opens up room for anthropomorphic bias in the evaluation of models. A qualitative inference requires judgment calls to be made, and a holistic inference to the best explanation requires that lots of information be considered. There are a lot of opportunities for bias to creep in. One might then point out that Morgan's Canon, despite its faults, is at least a decision rule that is applied objectively, preventing researchers from putting a finger on the scale for their favorite model.[10] As such, one might conclude, it is not clear how much of an improvement my suggested approach makes over Morgan's Canon.

I share the worry about bias in inference to the best explanation, and indeed it is part of the reason I made this my second challenge in the book. I have two things to say about it. Firstly, I think one of the main virtues I outlined in the last chapter remains; this procedure puts disagreements out in the open where they can be considered and discussed. If a researcher thinks that the existing evidence favors one hypothesis over another, they need to articulate why. As things stand, researchers can fall back too often on simply asserting parsimony claims, which won't do. I realize this is not a full solution, so, secondly, I suggest a much more robust project of studying human anthropomorphic bias. This would be helpful anyway, as I noted that bias can impact any of the stages of

---

[9] Nemati (2022) suggests that we can cordon off the use of anthropomorphism in the so-called "context of discovery," and expect that any biases will then be ironed out by experimental results in the "context of justification." While I remain open to strategic uses of anthropomorphism, the solution can't be this simple, as implicit anthropomorphism is just as much an issue in the context of justification as discovery.

[10] This argument form is made in discussions of psychology statistics, especially in light of prominent replication failures (see Chapter 6): the standard of p = 0.05 for statistical significance is arbitrary and has its critics (Machery 2021). But, even so, it prevents individual authors from choosing the value that engineers the result that they want (down the 'garden of forking paths,' e.g., Romero 2019).

the science. The goal, for present purposes, would be to develop interventions or methods that researchers can use to reduce the impact of anthropomorphic bias. Ideally, these could be applied at any stage, including an evidentialist evaluation of the models. With that setup in mind, I turn now to research on the human tendency to anthropomorphize.

2.4 The Psychological Study of Anthropomorphism

As a topic of study, the human tendency to anthropomorphize connects with several large research areas, including work on theory of mind and social cognition generally. Nonetheless, as a phenomenon itself, it has received surprisingly little direct study. Research has picked up considerably in the last few years. Things are moving in a promising direction, but we have a long way to go.[11]

For my purposes here, the results we currently have are informative but are mostly oblique to the specific question at hand: how might we control anthropomorphic bias in comparative psychology? Indeed, it is difficult to extract a cohesive research program on anthropomorphism because researchers are coming at it from different directions for different purposes. This is, to my mind, actually appropriate, because I don't think there is a single phenomenon of implicit anthropomorphism. Many different mechanisms in human folk psychology likely influence decisions in various ways. And the interesting effects will vary by context. Not all anthropomorphism is an error, and the study of the mechanisms involved is orthogonal to the question of whether anthropomorphic claims are accurate; but we are nonetheless, understandably, interested in the cases in which it can produce errors. For example, work on anthropomorphizing God is concerned with whether participants attribute physical properties to a spiritual entity (Barret and Keil 1996; Shaman, Saide, and Richert 2018), which clearly is not a concern with thinking about robots or animals. Along similar lines, it seems reasonable to consider there to be some kind of error in attributing a mind to a simple robot, but the kinds of errors we ought to worry about (as noted) are very different for a chimpanzee or an octopus. The worries there are not that we attribute a mind or even just that we attribute "too much" mind, but that we attribute the wrong sorts of minds, or

---

[11] I'll also note here that a better understanding of human implicit folk psychology and mind perception might help generate better models for the debate about animal mind-reading discussed in the last chapter.

that we miss minds that are there. Even within the subject of animals, we might expect different manifestations with different species or in the observation of behavior, as opposed to the design of experiments or the evaluation of models, and so forth.

This section can't provide any concrete solutions for the science of animal minds. Instead, I present work on the human tendency to anthropomorphize at two levels. First, I describe the most prominent theoretical framework for understanding anthropomorphism, the three-factor theory proposed by Epley, Waytz, and Cacioppo (2007), along with some basic findings that elaborate each of the three factors. Second, I survey the methods that have been applied in this research, with an eye towards productive future work. Work on anthropomorphism has been simultaneously quite creative in its methods and still overly reliant (for our purposes) on direct reports.

2.4.1 A Three-Factor Theory

Epley, Waytz, and Cacioppo (2007) provide the most systematic account of the psychology of anthropomorphism on offer. They note that there had been, at the time of their writing, very little research on the actual formation of anthropomorphic beliefs. As such, they base their account on understandings of how people think about other people. The core claim is that people use salient knowledge about minds they know well (their own minds or human minds more generally) and apply that to nonhuman entities. The authors identify three overarching factors that they take to influence the degree of anthropomorphism in any case. That is, these three factors make information about minds salient and so likely influence the tendency to apply that knowledge in interpreting nonhuman entities. The three factors are: elicited agent knowledge, effectance motivation, and sociality motivation. Each of these factors has individual dispositional, situational, developmental, and cultural elements. Each has also received some support from subsequent work. I will briefly survey these ideas here, though I cannot provide a complete picture of the state of knowledge.

Elicited agent knowledge is, in short, the amount of information about minds that is present to mind at the moment ("elicited agent information" might have been a more accurate term). In the present context, the most interesting example might be the 'similarity' of the animal to a human in looks or behavior. For instance, Eddy, Gallup, and Povinelli (1993) asked participants to rate thirty different animal species from across phylogenetic groups on two different scales. One ranked how "similar they are to you" (the actual language of the prompt), and the other ranked their ability to perform three cognitive tasks. The results showed strong correlation between the similarity scores (however people conceive those) and willingness to attribute cognitive abilities. Arico et al. (2011) summarize the crucial features for attribution of mind in three categories: having facial features (especially eyes, like Mr. Trash Wheel), displaying nonlinear movement trajectories, and displaying contingent interaction with the environment. In addition, Morewedge, Preston, and Wegner (2007) asked participants to rate the degree to which various entities were conscious and compared that to speed of movement. Across various categories of target entities, they found that entities that move at near average human speed were rated as more conscious than those slower or faster. This included a collection of nonhuman species, animated robots and objects, and even faster and slower humans.

Effectance motivation, the second factor, is the motivation to be effective in one's interactions with the world—to understand what is happening and how to engage with it. In this context, this need not mean actually trying to do something with the animal. The desire to understand, predict, and explain its behavior would still count. Waytz et al. (2010) tested this proposal in a few ways. They found that participants were more likely to anthropomorphize unpredictable computers, gadgets, and robots (and in one fMRI experiment found increased activity in areas of the brain associated with social cognition when anthropomorphic responses were high), on the thinking that unpredictability increases this motivation. In another experiment, they motivated participants by paying them to predict the behavior of a robot in a video. When so motivated, participants showed greater anthropomorphism. Finally, they also found that participants felt they had a better understanding of entities in a vignette when they were told to anthropomorphize than when told to view them objectively. This suggests that anthropomorphic explanations satisfy the effectance motivation.

Sociality motivation, the third factor, is the motivation to make social connections. Epley, Waytz, and Cacioopo (2007) mostly tie this to lacking connections and loneliness. Epley et al. (2008) found that tendency to anthropomorphize correlated with self-reports of loneliness and could be manipulated by inducing loneliness. However, loneliness is not the only reason one might want to form and maintain relationships, including those with animals that one studies.

So, we can think of possible specific factors that influence anthropomorphism as belonging in one of these three categories. The ways each will

60 ANTHROPOMORPHIC BIAS

manifest will vary context to context. Even so, I think that this can helpfully orient us in thinking about what interventions might influence implicit anthropomorphism.

2.4.2 Methods

Existing methods for studying anthropomorphism can be roughly divided into three categories. First are direct reports, in which participants rate the degree to which they are willing to attribute some humanlike or mental traits to the target of the study. Second are indirect reports, in which participants describe or explain behavior, and researchers look for anthropomorphic language and posits in that explanation. Third are studies that use implicit measures that might indicate anthropomorphic reactions to situations. I'll run through each of these, presenting some interesting examples and reflecting on the strengths and weaknesses of each.

The most common measure of anthropomorphism in the experiments described above was an explicit report; participants were asked to rate how willing they were to attribute a mind or beliefs to the entity in question, or to rate how much it seemed to have these properties. Indeed, this is arguably the dominant method. Among the most advanced versions of this approach is the individual difference in anthropomorphism questionnaire (IDAQ) (Waytz, Cacioppo, and Epley 2010). IDAQ is a thirty-item questionnaire that asks participants to respond on an eleven-point scale (from *not at all* to *very much*) to anthropomorphic and, as controls, non-anthropomorphic questions.[12] Anthropomorphic questions include "To what extent does the average mountain have free will?" and "To what extent does the average insect have a mind of its own?" Non-anthropomorphic questions include "To what extent is the average amphibian lethargic?" and "To what extent is the average robot good-looking?" They find that individuals' scores on the scale are reasonably stable and correlate with a number of traits and dispositions of interest; those more likely to anthropomorphize are more disapproving of destroying a computer or a flower garden, for instance. Ruijten et al. (2019) develop another

---

[12] It's worth noting, given my emphasis that anthropomorphism may vary with context, that Waytz, Cacioppo, and Epley originally included questions naming entities of four kinds—animals, natural objects, technological devices, and spiritual agents. An early factor analysis revealed that participants were not distinguishing the anthropomorphic from non-anthropomorphic questions about spiritual entities, suggesting that they were simply reporting their degree of belief in spiritual agents. As such, those items were removed from the questionnaire.

questionnaire-based measure. They argue that it improves on IDAQ because it probes a broader range of kinds of agents as well as kinds of anthropomorphic traits. Their measure also includes a ranking of items based on "difficulty" of making the anthropomorphic attribution, which allows more detailed modeling of responses.

In general, explicit report methods are useful because they are easy to run, and they allow control over a lot of different variables that might influence anthropomorphism; several features of the situation and the stimulus materials can be varied, and then subjects can simply be asked what they make of them. They also fit very well with the traditional understanding of anthropomorphism as positing one of a set of humanlike traits. They have also tended to assume that it means overestimating the intelligence of animals: they tend to ask whether some sophisticated trait is present or not, though this is not necessarily part of the method (e.g., one could ask whether a grinning macaque looks happy or afraid). So I take these to provide useful information, especially on coarse-grained features that can cause anthropomorphism, like those in the three-factor model. However, to understand the role of anthropomorphism in comparative psychology, it is best to also have methods that can isolate unconscious, automatic contributions.

The second approach is to ask participants to describe or explain a situation and code their language for anthropomorphic attributions. Likely, Heider and Simmel (1944) originated this method in a classic experiment. Heider and Simmel had participants watch an animation in which geometric shapes move around a screen and asked them to describe what they had watched. In the animation, the shapes move in animate-looking ways—stopping and starting, seemingly approaching and withdrawing. Indeed, the shapes appear to interact: a bigger triangle appears to bully the other shapes, which run and hide. Heider and Simmel found that participants regularly, spontaneously, and freely described the scene in terms that attributed agency and intention to the shapes. In another famous study, Barrett and Keil (1996) used another version of this kind of task. In their main experiment, participants listened to a short story in which God performs some action. The stories were followed by questions about whether the story included specific pieces of information (yes or no). Many questions probed information about God's role in the story that was not mentioned but would be inferred by anyone attempting to understand the story through an anthropomorphic God concept: for instance, needing time to physically move between two different locations. They also asked participants to fill out a questionnaire on their theological beliefs. They found that participants readily affirmed that information in the anthropomorphic prompts had been in the story despite the fact that 1) it was not and 2) they had extrapolated the information based on (implicit) assumptions about God that violated their explicit theological beliefs. Looking for anthropomorphism of dogs instead of gods, Horowitz and Bekoff (2007) videoed participants playing with their dogs in parks around New York City. In post-play questionnaires, they found that owners were more likely to attribute intentions and understanding to their dogs when the play had 'gone well' by the owners' assessment. They note four types of behavior that were repeated across successful play sessions: directed responses to the other, indications of intent, mutuality, and contingent activity.

This type of method may have some things going for it as a way to understand research into animal minds; after all, constructing descriptions and explanations of behaviors is exactly what researchers do. We might worry about whether the explanations that nonexpert participants produce would transfer to researchers who are bringing all their theoretical knowledge to bear on their explanations. Nonetheless, there may be patterns in this type of work that can illuminate general, implicit forces that even experts may not control as effectively as they think they do.

Lastly, some research has applied various methods for tracking automatic responses to anthropomorphizing cues. This class of methods is a bit of a grab bag, but there are many interesting ones. These methods have picked up significantly in the last few years, as researchers become more interested in the ways that people react emotionally to machines. This is part of research in social robotics but also more generally in marketing and other transactions humans might go through with machines (see for reviews Hortensius, Hekele, and Cross 2018; Chugunova and Sele 2020; and Harris and Anthis 2021).

Waytz, Heafner, and Epley (2014) ran a test in which participants in a driving simulator drove 'self-driving' cars. In this test, they measured heart rate and startle responses (independently coded from video) and combined them into a behavioral measure of trust in the car. Unsurprisingly, they found greater trust when the car was given more anthropomorphic features. Physiological markers of emotional response like these have been a common target. Rosenthal-von der Pütten et al. (2013) used skin conductance and electrocardiograms to track participants' emotional responses to videos including a robot dinosaur named Pleo. In one video Pleo was "tortured" (their word), while in the other Pleo was treated in a friendly manner. Participants, predictably, showed a greater physiological-emotional response to the torture video. They varied whether participants had been able to play with Pleo or not before watching the videos. However, this did not modulate the difference in responses. In questionnaires afterward, participants were more likely to attribute feelings to the robot and reported more empathetic concern after the torture video than the play video. Cross et al. (2019) had some participants socialize with a robot for five days before their crucial test, which involved participants in an fMRI machine watching videos of the robot "in pain." They found brain activity and subjective reports of similar levels of empathetic response between videos of the robot in pain and a human in pain. However, they did not find any effect of socializing. Menne and Lugrin (2017) tracked and coded facial expressions of participants watching similar videos with a different robot and were able to track differences in facial responses to videos of positive and negative treatment of the robot.

Different kinds of eye tracking have also been employed. Staudte and Crocker (2009) used eye tracking to see if participants followed a robot's "gaze" when attempting to understand its speech. Marchesi et al. (2021) tracked the size of participants' pupils while they observed images of a robot performing actions and assessed mechanistic as opposed to mentalistic descriptions of the action. The thinking is that pupil dilation can index some cognitive activities, corresponding with cognitive arousal or load, indicating interest, difficult problem-solving, and unexpected results.

Not all measures have to be physiological, though. Złotowski et al. (2018) tested implicit anthropomorphism using a priming paradigm in which participants briefly viewed an image of a robot before classifying a silhouette image as depicting a human or an inert object. They compared this to IDAQ as an explicit measure and attempted, but unfortunately failed, to independently manipulate the two. The attempted implicit manipulation had participants play a game with one of the robots used as a prime image, while manipulating its apparent emotionality. Li et al. (2022) used a variant of the Implicit Association Test which probed whether participants associated robots with high agency or high experience. They found that humans implicitly associate both traits less with robots than with humans (though seemingly associate robots more with agency than experience). Comparing results on this implicit measure to an explicit measure, they suggest that these might be related but independent processes. Finally, Spatola et al. (2019) note that people often display better attentional performance in the presence of other humans. They gave participants Stroop tests (of attentional control) to perform in the presence, or absence, of a humanoid robot. They found improvements in the presence of robots that triggered anthropomorphic attributions in earlier verbal interactions, which were similar to the improvements made in the presence of other humans.

These implicit measures, and others like them, hold intriguing promise. They have emerged only in the last few years, so there is much to do. Implicit measures in general can be difficult to interpret, and can't be guaranteed to correlate with behavior in the world. Even so, they provide new windows on the various forms anthropomorphism can take.

2.5 Implicit Anthropomorphism and Affective Priming

Until recently, explicit measures had been the dominant method in the empirical literature on anthropomorphism. Explicit measures have two main limitations based on my arguments here. First, they do not probe the implicit forms of anthropomorphism, which I have argued are actually the core worry in comparative psychology. Second, they tend to fit the standard view of anthropomorphism described above: to anthropomorphize is to posit the presence of particular 'humanlike' traits which are absent (or, at least, may be absent), and this typically means inflating their perceived intelligence or sophistication. The recent arrival of implicit methods in work on robotics helps considerably with the first problem. Arguably, though, we still lack a reliable and well-validated implicit measure (Złotowski et al. 2018). Plus, even the implicit measures listed above, I suggest, still only probe anthropomorphism as it manifests with the positing of particular traits. In general, these involve positing experience or agency, or treating a robot as if it had emotions or some sort of social agency. These are important questions for social robotics but do not apply as directly to nonhuman animals. The question with many nonhuman animals is not whether they have emotions but what sorts of emotions they have.

With the goal of developing an implicit measure that can avoid both worries, Jen Coane and I (Dacey and Coane 2023) have worked to develop a measure of anthropomorphism based on an affective priming paradigm developed by Carroll and Young (2005). This measure tracks unconscious aspects of (apparent) emotion recognition. Carroll and Young primed participants with images of human faces making expressions matching one of five basic emotional categories: happiness, sadness, fear, anger, or disgust. Participants were then asked to categorize words as belonging to one of these categories. They categorized the words more quickly and made fewer errors when the prime and target categories matched. For example, when presented with a happy face, subjects would categorize the word "merry" as relating to happiness more quickly and reliably than they would categorize the word "petrified" as relating to fear. The onset of the prime only preceded the onset of the target stimulus by 750 milliseconds, which is generally acknowledged to be too fast for conscious or strategic processing (Neely 1977). The implication is that, within that three quarters of a second, participants' automatic processes identified the facial expression as an instance of a particular emotion and began unconsciously activating other processing relating to that emotional category.

Our experiments attempt to leverage this effect. To simplify, we reduced the categories to 'happy' and 'sad', so that participants only had to track two categories. We piloted several pictures of animals that looked happy, sad, or neutral, and chose those that received the most consistent (explicit) ratings by participants in each category. We ran four experiments. Our first experiment included animals from all over the kingdom, looking happy or sad in different ways: all kinds of animals (not just primates) can look happy or sad for reasons that may not correlate with their actual mental state, including facial expression, body posture, and even the angle from which the photo is taken. In this version, we found priming in older adults, but not younger. In two other experiments, we used primate faces on uniform backgrounds (to reduce variance) and human faces (as controls). In both cases, we found priming effects as expected. Finally, we added interventions to speed up responses (an imposed deadline) and slow down responses (degraded, harder-to-read target stimuli). The priming effect disappeared when responses were sped up. Somewhat curiously, across experiments the results only showed up in measurements of response accuracy and not response time. It also appeared that, with nonhuman primes, the effect was mostly driven by negative apparent emotions. In three of the experiments, we gave participants IDAQ questionnaires but found no correlation between individual IDAQ scores and priming effect sizes. We did find that older adults in the 'all taxa' version of the experiment (and only that version) showed greater priming than the younger adults but lower scores on IDAQ questionnaires that we administered concurrently.

So at least some of the time, animals do *look* happy or sad both on conscious assessment (in the pilot study) and to unconscious systems (in the experiment itself), even if the cues people pick up on have nothing to do with happiness in that species. In any case, the main goal at this point is to develop the measure. Most importantly, this does demonstrate a possible measure that avoids the two worries listed above for existing measures: it tests implicit rather than explicit anthropomorphism, and it does not only test for the inflation of perceived intelligence or the attribution/nonattribution of some set of 'humanlike' traits. In principle, this measure could be paired with various interventions that might increase or decrease anthropomorphic tendencies, to see what works and doesn't work. It may also provide ways to test dissociations between implicit

66 ANTHROPOMORPHIC BIAS

and explicit anthropomorphism: the comparison between IDAQ and priming effects in the 'all taxa' experiment provides some hints worth exploring.

The hope, overall, is for a generally useful measure of implicit anthropomorphism that can be paired with various manipulations to better understand how it works. IDAQ has served this purpose well for explicit anthropomorphism, but there is no analogue for implicit anthropomorphism. If we have such an implicit measure, we could, for example, test the interventions that Epley and colleagues used to support their three-factor theory (Section 2.4.1). We could also test whether prompting subjects with a statement like Morgan's Canon influences implicit anthropomorphism. Debiasing techniques developed in social psychology also provide interesting interventions to test. Our affective priming paradigm has some promise for this role, but perhaps some other measure (one discussed in the last section or not yet devised) will prove better. In any case, this is the project that we need to pursue to get a strong grounding on how implicit anthropomorphism works, and how to control it.[13]

2.6 The Big Picture

From one point of view, anthropomorphism is a narrow topic: a particular error (or set of errors) that occurs when considering a particular type of question about a particular subject matter. However, if we step back to look at the bigger picture, it reaches into some of the biggest questions in philosophy and science.

For starters, it connects to questions about what kinds of evidence and inference we find acceptable in science. Specifically, it instantiates debates about objectivity in philosophy of science more generally. Opponents of anthropomorphism, for example, sometimes treat it as a deviation from proper standards of scientific objectivity—letting one's feelings, hopes, or desires drive one's conclusions. Even if we err when we deny anthropomorphic traits to animals, this line of thinking implies, at least we err while applying 'proper' reasoning practices. Correspondingly, anthropomorphic error represents a deeper mistake, because it results from a failure to do the right kind of thinking (Sober 2005, e.g., notes this line of thought). Thus, of the two kinds of error discussed in Section 2.3 (and Figure 2.1)—anthropomorphism and

---

[13] Limiting anthropomorphism would likely be the goal for researchers in comparative psychology, but there might be many other areas where it is fruitful to encourage it, including social robotics and policy discussions relating to animal welfare.

anthropodenial—anthropomorphism is the deeper mistake. Defenders of anthropomorphism (or, at least, those less worried by it) might note in reply that human values are ineliminable from scientific thinking (Douglas 2000). Thus, the argument goes, the purportedly 'objective' stance of preferring errors of anthropodenial *is* a choice of a certain value—presumably in this case valuing objectivity (under a certain conception). But once we recognize this, we might note that moral values tell in the opposite direction. The psychological abilities we attribute to animals will impact the way they are treated, so underestimating their intelligence and their experience may lead us to treat them worse than our considered ethical positions dictate. As such, one might argue, we should prefer *overestimating* sophistication as a caution against doing harm. Regardless of which side one takes, it should be clear that the choice of values is necessary here, and it cannot simply be taken for granted that a certain conception of objectivity does not itself represent a value choice.

Anthropomorphism is also a core issue for questions about how we relate to animals, scientifically, ethically, and personally. Do personal, empathetic connections with animals encourage too much, or too little, caution in how we make various choices that impact how they are treated (perhaps most importantly decisions about what we eat)? What specific influence does the way we perceive, or deny, the humanlikeness of animals have over big-picture questions about our position and role as human beings in the world? On a more day-to-day basis we might wonder whether anthropomorphic conceptions of animals we interact with regularly get it right, or represent some motivated error: are our emotional bonds with our pets really real or just one-directional illusions? On the scientific side, should researchers form relationships with their animal subjects or view them dispassionately as target systems of study?

In considering questions like these, work on animal minds has been a prime target for arguments in favor of more expansive conceptions of reason in philosophy and science. These include arguments for the value of emotion and personal connection in understanding others (Gruen 2015) and the importance of poetry over argument as a mode of philosophical exploration (Diamond 2006; Mulhall 2009). In these discussions, work on animal minds has obvious (narrower) methodological interest but also serves as a valuable microcosm of discussions about how we think of ourselves and our place in the world.

Those inclined to such expansive views of philosophical reasoning might find my approach somewhat reductive—emphasizing the possibility of error in intuitive folk psychology, enumerating cognitive biases, and calling for more science to help settle the core issues. Advocates for objectivity, on the other hand, might think that I am being overly permissive by allowing any role at

68 ANTHROPOMORPHIC BIAS

all for unconscious, possibly affective, psychological processes in scientific reasoning. I think either reaction would miss the core of what I am trying to do, but there is a reasonable sense in which the approach is a middle ground on the spectrum here (recognizing that most individuals likely don't fall on either extreme anyway). There is no such thing as a truly dispassionate, disconnected reason, and we should not want it in this case even if we could have it. Even so, we must hold ourselves to standards, and determining what those standards are requires close attention to the details of anthropomorphic bias. Really, this is another instance of the balancing act required when we recognize and attempt to manage the fact that much of our cognition is not under direct control. The standards we apply are as much at issue as the results we argue for.[14] This works out in two key ways in the context of controlling anthropomorphic bias in the science of animal minds.

First, the goal here cannot simply be to eliminate activity by the processes of intuitive folk psychology. As I have said, when our folk psychology fails to pick up on cues, we can *miss* intelligence that is there. To the extent that we rely on intuitive thinking at various stages of the scientific process, and I don't think we have much choice other than to do so (we are human after all), the goal should not always be to *suppress* these factors but to try to *tune them* appropriately. What this means may vary by research question, species of focus, research stage, and so forth. There is not a single general solution here, as each of the constellation of anthropomorphic biases will manifest and operate differently in different contexts. For this reason, gaining the appropriate understanding of anthropomorphism will require a plurality of methods but also intentional, value-influenced decisions about how to tune the activity of these processes.

Second, part of this difficulty is that we cannot actually tell which specific judgments driven by unconscious folk-psychological systems are erroneous, and which are not, before we enter into a research project. To do so, we would need to know how animal minds work, and that is precisely what we are trying to find out. In comparison, consider work on other cognitive and social biases. As surprising as the effects discovered by that work can be, the errors that people make can be identified by comparison to some set standard: egalitarianism in the case of social biases and some rationality norm in cognitive biases (often the axioms of probability). So, even as we look for the appropriate tuning

---

[14] For comparison, I see a shared spirit with work on agency in light of unconscious processing (e.g., Doris 2015); instead of revising standards of agency, the goal here is reconsidering standards of rationality (but see also Stanovich and West 2000).

of our folk-psychological capacities in a given context, we cannot simply check against an established standard. The tuning process will need judgments of appropriateness that pull from many sources.

So implicit folk psychology is not merely an ineliminable part of our reasoning about animal minds; it is an important one as well. Even so, it is something that we need to be aware of, study more, and recognize as a potential source of error. I think it's reasonable to view this 'tuning' as striking a balance between calls for more expansive conceptions of reason and calls for objectivity. In this sense, implicit anthropomorphism is a microcosm for unconscious processes in general. At first, we might find their influence frightening or problematic, but what we must do is learn to live with them in a healthy, productive way.

2.7 Conclusion

Work on anthropomorphism, and the actual impact it has on the science of animal minds, is ongoing. Even so, researchers should remain modest about the fact that implicit anthropomorphism may impact their reasoning, and should do so in ways that we do not understand well. More concretely, we can take two main points away from this discussion.

First, anthropomorphism, in the form that does and should worry researchers on animal minds, is a cognitive bias. This means that it is not merely a tendency to posit some list of 'humanlike' traits to animals, and it does not merely involve inflating the perceived intelligence of animals. It can lead to errors that do not impact perceived intelligence (as in attribution of the wrong emotions), and it can lead to missing intelligent actions that don't fit the expectations of our intuitive folk psychology.

Second, in order to truly understand the impact that anthropomorphism has on the science, we need to study it empirically. We cannot simply trust our ability to intuit what anthropomorphism might do in any given case. Researching this will require the development of new empirical measures, but there are several promising options being developed.

Where, exactly, this research program takes us is not clear. But, in the meantime, we can consider some tentative conclusions. For one, anthropomorphism need not be seen as the bogeyman that neobehaviorists see (Kennedy 1992). It is just one among many peculiarities of human reasoning. It is also not determinative: it doesn't force a certain conclusion, but instead might make some hypotheses or interpretations feel a bit more plausible. We should think of it as more of a nudge than a decisive judge. A greater fear might be reasonable if anthropomorphism were truly a consistent, uniform pressure in a single direction, as it is been historically seen. But I don't think it is; in different contexts, anthropomorphism will push in different directions. Moreover, there's good reason to think these tendencies can be fragile. We should be wary of feedback loops, of course, but it's not clear these are consistently happening in the field. On the other end of the spectrum, advocates for anthropomorphic methods may be too cavalier. There is little reason to think that we know how to limit anthropomorphic bias when engaging in anthropomorphic methods that likely encourage it.

While we wait for these results, there are some things we can do. Soll, Milkman, and Payne (2015) summarize strategies for debiasing that come out of the literature of cognitive biases more generally. Some of these suggestions may be helpful (different ones in different contexts, so I won't summarize here). Being presented with counter-stereotypical examples is also an important tool for reducing social biases (Dacey 2017a). One salient stereotype here would be that all intelligence operates like human intelligence. To counter this stereotype, one might consider strange or alien-seeming examples of intelligence, such as those presented by octopuses and bees. This advice is generic, and may or may not work, but it seems worth trying.

I close by pointing out that the most important things to do are things that I suggest doing anyway. As suggested in the last chapter, make the reasoning for why an experiment is supposed to support some model of cognition explicit and public, where it can be evaluated. Carve up big questions like "do chimpanzees mind-read?" or "do rats reason causally?" into smaller questions, through the strategies mentioned there. These smaller questions will be less loaded and perhaps less likely to activate anthropomorphic responses. Also, as discussed in the next chapter, get more precise about the models used; models that simply name folk-psychological mental states will be especially prone to anthropomorphism. Models that specify concrete predictions will be easier to evaluate. But that, as we will shortly see, requires considerable work.

3

Modeling

3.1 Introduction

Up until the middle of the nineteenth century, magnetism, the electromagnetic force, and light each remained among the deepest mysteries of the natural world. Then, James Clerk Maxwell published his famous set of equations, which described electric and magnetic phenomena in a single set of terms and predicted light as a propagating electromagnetic wave. With these equations, science finally got a grasp on these deeply mysterious phenomena. A similar landmark occurred in neuroscience at the end of the nineteenth century when Santiago Ramón y Cajal was able to stain neural cells and draw detailed images of what he saw under the microscope. These drawings depicted the shape of neurons and showed that they were indeed distinct cells. Lastly, in 1957, Noam Chomsky's *Syntactic Structures* presented one of "cognitive" science's first great successes by demonstrating the potential for terminology and formalisms taken from computer science to explain linguistic phenomena that the then-dominant behaviorist paradigm could not (Chomsky 1957).

In each of these cases, science advanced because of a new model. Each takes a different form: Maxwell's equations, Ramón y Cajal's drawings, and Chomsky's model computing systems. But they all provide new tools for thinking about the target phenomena that allowed researchers to grasp them (or some portion of them), generate new predictions, and evaluate evidence against those predictions (Weisberg 2013).

Unfortunately, models in comparative psychology are often too imprecise to generate specific predictions. These models are often verbally described and massively underspecified in causal detail and predictions that they can generate. It is, however, not always easy to know how to do better. This is the core challenge of the chapter: it is hard to model the mind in terms that are precise enough to generate specific predictions.

Authors have raised concerns about models in comparative psychology (e.g., Buckner 2011; Penn 2011; Heyes 2015; Mikhalevich 2017) and psychology more generally (e.g., Luce 1995; Yarkoni 2022) for various related

Seven Challenges for the Science of Animal Minds. Mike Dacey, Oxford University Press. © Mike Dacey 2025.
DOI: 10.1093/9780198928102.003.0004

72 MODELING

reasons. Perhaps most forcefully, Allen (2014) argues that the lack of mathematical models in comparative psychology has been a major impediment to progress. The lack of specificity of cognitive models is also a significant contributor to the challenges of the first two chapters: it is a major source of underdetermination (Chapter 1), and the fact that these models are often filled in using intuitive, folk-psychological categories is a significant source of worries about anthropomorphism (Chapter 2).

It is extraordinarily difficult to describe the operations of the mind in terms that are precise, predictive, and illuminating. We struggle to understand what the mind really *is* in large part because we lack the terms (verbal, mathematical, or otherwise) to describe exactly what it does. As such, the development of new types of model of the mind has frequently come with proclamations of revolution (whether or not actual revolutions have followed). Most significantly, the cognitive revolution of the mid-twentieth century (Chomsky included) brought with it concepts of information and computation that still dominate descriptions of the mind today. More recent proclamations of revolution have attended the development of connectionism (Churchland 1981; McClelland and Rumelhart 1987), dynamical systems theory (Chemero and Silberstein 2008), multilevel mechanistic explanation (Boone and Piccinini 2016), and Bayesian or predictive processing (Clark 2015). Setting aside the merits of any of these in particular, the general lesson is that it is tremendously important to develop new modeling tools where we can; or, as will be the strategy here, to deploy old tools in new ways to produce something functionally new. But there are more specific problems in comparative psychology as well.

Suppose, for example, a rat sees a light in an experimental chamber and shuffles over to the hopper, anticipating that food will be released. Clearly, it recognizes some connection between the light and the delivery of food, but what kind of connection? Does the rat recognize a causal relationship between the events, or is it merely the case that the light automatically makes it think of food? In the terms it is usually cast, is the rat capable of causal reasoning, or is it simply operating on associations? Presumably, these two ways of comprehending the situation would entail different abilities to engage with the world; causal reasoning is generally thought to be significantly more powerful than association. However, when an ability like causal reasoning is simply labeled with this verbal expression, it doesn't tell us much about the process involved. What specific abilities does it allow? When and how are these abilities engaged? Verbally expressed cognitive models often cannot answer these questions.

Much of comparative psychology breaks down along an associative/cognitive divide—an associative explanation pitted against some more sophisticated "cognitive" explanation. The literature on causal reasoning in rats makes a good example for our purposes because the candidates are in some ways similar: these are each ways of apprehending sequences of events. Associative processes, as they are commonly understood, link mental representations in rigid sequences based on simple relations, like prior pairing in experience (e.g., conditioning). Cognitive models include, in effect, anything more sophisticated: mind-reading, most forms of behavior-reading, mental maps and models, and so forth. Causal reasoning in particular involves treating a sequence as causal, which allows various ways of reasoning about it (discussed below). These two model types are taken to exclude one another; in the rat example, if we predict behavior with an associative model, it is taken to be an associative process. If the behavior is too complex to be associative, it is taken to be causal cognition. Cognitive models are typically characterized verbally, as I just listed them. Associative models, on the other hand, are often mathematically characterized.

The associative/cognitive divide has been at the heart of the problem of modeling in comparative psychology (Penn and Povinelli 2007b; Buckner 2011, 2017; Hanus 2016; Dacey 2016b). It is also, I think, the place to look for one important solution. Specifically, I have argued for a reinterpretation of associative models that takes them to be partial, abstract descriptions of a process (Dacey 2016b, 2017b, 2019b). This means that they are no longer incompatible with cognitive models. As such, the mathematical precision of associative models can be applied in places it would not be otherwise. Associative models are valuable tools for describing phenomena in which the sequence of states is interesting or important, regardless of the type or complexity of process involved. Before I can explain exactly how this addresses the problems, I set up the challenge of modeling cognitive processes, and then the role that the association/cognition dichotomy plays.

3.2 The Trouble with Models in Comparative Psychology

Allen (2014) provides the strongest arguments against the current state of models in comparative psychology: "Put bluntly, there is too much trophy hunting and not enough theory" (76). His worry is that cognitive models are often simply presented verbally. Instead, he argues, we need mathematically precise models that can generate specific predictions to test. Lacking precise predictions like these, researchers are left to "trophy-hunt"—to simply find some impressive-looking behavior they can get an animal to display.

74 MODELING

The verbal formulation of models has been a frequent target of criticism. For example, Penn (2011) argues that verbally expressed hypotheses in comparative psychology are usually filled in by intuitive folk psychology, which exacerbates anthropomorphism. A researcher looking to generate a prediction from a general hypothesis like "this species has causal cognition" might simply intuit something that seems reasonable. This reliance on implicit folk psychology opens the door for implicit anthropomorphism.[1] Discussing psychological work on humans (rather than animals), Yarkoni (2022) argues that verbally expressed models are taken as grounds to unjustifiably generalize experimental results beyond the hyperspecific (usually artificial) context of that experiment. For example, the idea that a particular experiment provides evidence for causal cognition (my example) is taken to imply that the capacity, generally cast, is present, which in turn justifies generalization to other situations where the capacity might have an impact. This generalization is not actually justified because the effect might be a fluke of the situation, rather than a result of the general capacity. He takes this to be a major contributor to the crisis of failures to replicate across psychology (he calls it a crisis of *generalizability* rather than *replicability*, as it is usually framed; more on replication in Chapter 6).

Though the precise worries here differ, they share their focus on the fact that cognitive models are not well specified. Saying, for instance, that an animal possesses a capacity for causal reasoning does not tell us how that capacity operates, when it will operate, or what effects it will have.

When contrasting causal reasoning with association specifically, there are many candidate differences that have come up in various parts of the literature. One possibility is the recognition of causal sufficiency—that one cause can be sufficient for bringing about an effect. Others include causes occur before effects, while associations can be simultaneous. Reasoning allows the ability to think about possible causes that are absent at the moment. Causal reasoning might involve a recognition of difference between events that are merely observed and those that are caused by one's own actions. Causal reasoning may allow (and require) the recognition of some mechanism by which one event causes another, and indeed causal learning might require that such a mechanism be observable, at a minimum, typically requiring contact. Causal understanding can also shape expectations based on known properties of the objects involved: an object known to be heavy will be expected to initiate different events than an object known to be light. There are also innumerable specific

---

[1] Note again that this form of anthropomorphism need not produce inflated perceptions of animal minds; it might lead us to miss forms of intelligence that we are not looking for (Chapter 2).

interactions that could be recognized as causal, including pushings, pullings, crushings, and so on (Cartwright 2004). Even further, there are many different domains in which causal reasoning may or may not be employed, which follow different rules: fluid dynamics, rigid-body dynamics, social interactions, gravity (falling and support), weight, flight, wind, optics (diffraction through the water surface), and so on.

A species' ability to reason about causal systems may include or exclude any of these in any pattern. So, there is a live question how to draw the line (or lines) for causal reasoning in this multidimensional space; which sets of abilities in which domains would we require before we count a capacity as a capacity for causal reasoning. In the meantime, this is why verbally expressed models are so hard to test. If we test for one of these abilities and do not find it, that doesn't mean that causal reasoning, in some other form, is not present. And if we test for one and do find it, it doesn't necessitate that the others are present as well. This is how the lack of specificity in models exacerbates the problem of under-determination (e.g., Chapter 1).

To be clear, causal learning is among the better domains for formal/mathematical modeling. For instance, some authors model causal cognition as the representation and manipulation of causal graphical models, which come with formal rules for their use (e.g., Glymour 2001; Danks 2014). The actual learning of these graphical causal models has been modeled as a complicated statistical problem rather than one that makes use of particular understandings of particular kinds of systems (Gopnik et al. 2004). Thus, this is usually seen as a sort of domain-general causal reasoning mechanism. So even with this precision, and even if the basic approach here is right some of the time (i.e., human deliberate causal reasoning, arguably), there are a number of possible kinds of causal reasoning in the animal kingdom this sort of model might miss.

There are many factors that contribute to this general state of affairs. Firstly, as I have noted, modeling is hard. It not only requires the technical and computational skill to generate mathematical models but also requires coming up with the right sorts of models to test. Computational models require lots of assumptions, and face a "garbage-in, garbage-out" problem if one gets the assumptions wrong. It also often makes research opaque to those untrained in a specific kind of model, at the risk of fragmenting the literature. Allen (2014) further notes that some resistance to mathematical modeling likely arises because of the subfields that have employed mathematical models most; mathematical models tend to come either from neuroscience or from experimental traditions that have their origins in behaviorism (this includes the rat examples

76 MODELING

I rely on here). Most pessimistically, this might make one suspect that mathematical modeling amounts to an elimination of the mental. Rather than talk about mental entities such as reasoning and perception, these models describe neurons or patterns of stimulus and response. There have also been prominent eliminativist movements attached to each of these subfields (eliminative materialism on the neuroscience side, and behaviorism on the behavioral). However, there is nothing in these modeling strategies that requires elimination of mental entities. One might still worry, less pessimistically, that mathematization comes at a cost of actual understanding. Folk-psychological terminology allows us to actually grasp claims about the mind, which may be important for the progress of science but also for the actual understanding of our relationship with animals and for the communication of scientific results as they influence practical considerations such as animal welfare (e.g., Bekoff 2000).

Arguably, though, the main single cause of problems with models is an overemphasis on the dichotomy between associative and cognitive models. Smith, Cuchman, and Beran (2014) argue that focus on the question "is it associative or is it cognitive?" has led researchers to ignore more interesting questions, such as what information is processed and what it means to the animal. Several others target this distinction specifically, including Papineu and Heyes (2006), Penn and Povinelli (2007b), Buckner (2011), and myself (Dacey 2016b). Broader criticisms of modeling practice, I suggest, also amount to criticisms of the associative/cognitive distinction. Heyes (2015) argues that well-understood models of phenomena are often ignored in favor of the researcher's favored models (often an associative model vs. a cognitive model). Mikhalevich (2017) argues that the emphasis on models positing simpler processes such as association (due to Morgan's Canon) has fostered further development of these models at the expense of models of more complicated cognitive systems, in a self-feeding spiral. Though not all of them frame it this way, the general lesson I take from these critics is that the emphasis on associative versus cognitive explanations has led to neglect of other strategies that might generate new, more fruitful models.

Overemphasis on the associative/cognitive divide also produces a large gap between the models that are on the table. Associative models are thought to describe very simple processes, while cognitive models are often framed around human capacities (and not merely any human capacities, usually the most sophisticated). This means that we are left with severe difficulties in describing the processes that fall in the large gap between these two extremes of sophistication (e.g., Buckner 2019). While cognitive models are often underspecified, we completely lack models of most of these intermediate processes. Realistically, this is where most of the action probably is.

This framing is also partly responsible for the prevalence of "trophy hunting" that Allen describes. In one of his more evocative passages, Allen says: "To continue the pre-Newtonian analogy [he might have said pre-Maxwellian to fit our opening example], it is as if comparative cognition is presently in an alchemical stage- too much emphasis on transmutation of the animal, and not enough understanding of which properties are worthy of systematic investigation" (86). This occurs with the "associative versus cognitive" framing because the work of the purported cognitive process is often anchored against the associative alternative. In cases of trophy hunting, one is not really testing a cognitive model, one is simply trying to demonstrate behavior that would rule out any plausible associative explanation. Often, this means trying to demonstrate the smartest-seeming behavior that one can. In these cases, there doesn't need to be a prediction coming from any specific model, but the "associative versus cognitive" dynamic still motivates the project.[2]

There are likely many routes forward here, and, indeed, this is an area where I think we are best-off if a thousand flowers bloom. For example, Starzak and Gray (2021) propose their dimensional model of causal reasoning (discussed in Chapter 1) to get out of just this dichotomy in causal reasoning debates. This approach seems potentially fruitful and should be explored. Here, though, I develop another, non-exclusive, option: the reinterpretation of associative models. First, I have to clarify how associative models *are* interpreted in ways that contribute to the issue here. That requires a sense of how debates proceed.

3.3 The Tale of Causal Reasoning in Rats

The dichotomous treatment of associative models and cognitive models presupposes a certain interpretation of those models. To see how, it helps to see the dichotomy in action. So, in this section I describe a brief history of some work on causal reasoning in rats. This will also do double duty by demonstrating

---

[2] Trophy hunting has (at least) significant overlap with "success testing," as mentioned in Chapter 1. There, I supported moving to a "signature testing" framing. Nonetheless, Allen's alchemy comparison suggests reasons this "trophy hunting/success testing" strategy might make sense. He describes a science lacking a paradigm—what Kuhn (1962) calls "extraordinary science." In periods of extraordinary science, one of the things that researchers do is "cast about" experimentally, just to see what happens. Trophy hunting is, arguably, just this kind of casting about. Especially since so much research on animal minds before the last few decades presupposed them to be simple, there may be value to "casting about" at the upper ends of animal cognitive abilities. Even so, we need to be clear about what these studies mean, and we should aspire to move to a more model-driven science.

78 MODELING

how a lack of specificity in what is required to count as an instance of causal reasoning or association has played out in practical terms; as new effects are discovered and new models are produced, the types of processes that might be counted as members of each category have shifted, and purported requirements for causal reasoning have been abandoned.

The history has been one of slow (perhaps begrudging) increase in the complexity of the processes described by associative models. This has occurred through an iterated process whereby a new effect is taken as evidence of causal learning but then subsumed by a new associative model (this has played out in many domains, not just causal reasoning). All along, though, association has represented the simplest option (simplicity is, in many senses, the defining characteristic of how association is thought about; Dacey 2019a).

Our story begins in the behaviorist period;[3] we start with the simple conditioning-style experiment mentioned in the introductory section (3.1). A rat sees a light and expects food. This response could be explained by either type of mental ability. The associative story would be that each time the light is paired with presentation of food, there was a connection strengthened between the light and food. This is classical conditioning as Pavlov observed in his dogs. In behaviorist versions, the association is formed between the *stimulus* of the light and the *stimulus* of food. Since the food stimulus is rewarding, it can encourage the association to form. This view of association has largely been rejected in favor of views whereby the association is formed between internal, mental representations of these events/objects. Nonetheless, it likely still influences the way people think about associations (as Smith [2000] puts it, you should not be "[a]voiding associations when it's behaviorism you really hate"). Associative models of some form or other have consistently remained as a "permanent contender" (Hanus 2016) in debates about how animals perform some task. Indeed, because of Morgan's Canon, associative models are often treated as the *default* that must be ruled out in order for a more sophisticated model to be accepted (as described in Chapter 1, but in the context of causal reasoning see Penn and Povinelli 2007b; Hanus 2016; Starzak and Gray 2021; Halina forthcoming).

A major step in the development of associative models came with the Rescorla–Wagner (RW) model (Rescorla and Wagner 1972).[4] The RW model describes a process whereby the presentation of the light (in the case at hand)

---

[3] The history here of conceptions of association does not begin with behaviorism, of course (Dacey 2020, 2022b), but the behaviorist reading sets the backdrop for comparative psychology.
[4] See Soto et al. 2023 for more on the legacy of the RW model.

causes an expectation of food. The strength of the connection between a representation of the light and a representation of the food will be increased or decreased depending on whether the 'prediction' is borne out or not. This model represented a step forward because it was able to explain "forward blocking," whereby an existing association could block the formation of new associations (Kamin 1969). For example, if the rat has associated a light with food, and then a tone is played along with the light before presentation of food, the rat will not expect food when played the tone alone. This might be thought to indicate an understanding of causal sufficiency—that the light is a sufficient cause of food, so the tone can be ignored. However, the RW model shows that this need not be the case; the effect can be predicted using a model that posits only elements that are sufficiently simple to count as 'associative' (as standardly understood).

The RW model, however, does not predict a related effect known as backward blocking (Miller and Matute 1996). In this case, the rat is initially trained with the light and tone together, followed by food. After learning this association, the rat is shown only the light, followed by food. It might show some suppressed reaction to the light at first, but will learn to respond to it. However, if *then* played the tone, the rat will not react as if expecting food. Here, we might think that the rat, upon recognizing the pairing of the light alone with food will recognize it as the "true cause," and recognize the tone as redundant. The RW model does not predict this because the tone is not present at the time when it is blocked, and the RW model can only predict modification of associations when the stimuli are present.

This was initially captured by new associative models that posit associations between the light and tone (Van Hamme and Wasserman 1994; Dickinson and Burke 1996). However, again, backward blocking also seems possible when there is no direct association between the light and the tone (Denniston et al. 2003). And again, a new model called the extended comparator model was generated to predict this result as well (Stout and Miller 2007). In this model, existing associations compete not just at learning but also at use. The extended comparator model is much more powerful in the range of effects it predicts than earlier models. And the fact that it requires some kind of recall of other possibilities at use might make it something of an edge case for whether it counts as a model of an associative process or something more complex (Penn and Povinelli 2007b), but it is generally seen as an associative model.

Another important test for causal reasoning targets the ability of rats to recognize the difference between events that are observed and those that result from interventions. This is an important distinction in some conceptions of causal reasoning (e.g., those that cast it as the manipulation of causal graphical

80 MODELING

models). For example, a drop in atmospheric pressure is a common cause of both a change in a barometer reading and a storm. As such, we observe regular co-occurrence of the barometer change and the storm. If one were a mere associative learner, they would form a direct association between the two. When such a creature then reached into the barometer and lowered the reading, they would expect a storm. In contrast, an actual understanding of the causal structure tells us that our intervention on the barometer has no impact on the storm; the storm is caused by dropping atmospheric pressure, and in this case the barometer change was caused by intervention, not atmospheric pressure.

Blaisdell et al. (2006) (also Leising et al. 2008) tested whether rats can make this kind of inference. They trained rats in a chamber in which a light preceded both a tone and the release of food, establishing it as a "common cause" of the two effects. In this setting, if the light is covered, the tone can be a signal for the food, so a rat would reasonably expect food on the basis of hearing the tone. However, the experimenters also gave the rats the ability to initiate the tone by pressing a lever. The rat pressing a lever here is supposed to be like a human who intervenes on a barometer reading. The question is what happens when they press the lever while the light is covered so that they don't know whether it's lit (see Figure 3.1). If they understand the difference between observation and intervention, and the basics of the causal structure, they should know not to look for food when the tone is caused by their own intervention. Blaisdell et al. report significantly less searching for food after tones caused by intervention. They take this to be strong evidence of causal reasoning in rats.

Predictably, if you have been following along, Dwyer, Starns, and Honey (2009) propose a modified associative model, including response competition, which they take to explain the effect, and thus conclude that the process is associative rather than cognitive[5] (but see Blaisdell and Waldmann 2012 for reply). This process can iterate indefinitely and occurs across comparative psychology (in Dacey 2016b I describe the same process in the study of pigeon social learning). Eventually, as associative models seem to explain more and more, it becomes hard to tell what the actual distinction is (Hanus 2016; Buckner 2017). This is exacerbated by the fact that associative processes can be nested within other more sophisticated processes, and can include cognitive elements (Dickinson 2012). Even so, the associative/cognitive divide remains in much of comparative psychology and drives several of the problems described in the last section.

---

[5] They also note that they failed to replicate the result, so it's hard to know what to make of this overall. My focus here is on the dynamics of the debate rather than the right conclusion.

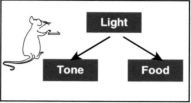

Fig. 3.1 A causal graphical model of the Blaisdell et al. (2006) experiment, in which the light acts as common cause of both the tone and food. Intervention causing the tone "breaks" (renders irrelevant) any other causal connections to it (namely, the light). From Blaisdell & Waldmann 2012.

3.4 Interpreting Associative Models (Properly)

The dynamic above is predicated on the idea that associative and cognitive explanations are mutually exclusive. That idea makes sense on the widespread assumption that each denotes a separate class of processes. Associative models are generally taken to denote the class of *associative processes*. This means that the application of an associative model to some process implies that the process is a member of that class. Associative processes are thought to be characterized by their simplicity and lack of flexibility. Associative learning processes are driven by mere repetition. Mental representations that are frequently co-presented will form an association between them such that, subsequently, the latter representation will become active once the former is presented.

The idea of associative processing arguably goes back to Hume, though its role in the associationism of the eighteenth and nineteenth centuries is debatable (Dacey 2015, 2022b). In the animal cognition literature, it is more commonly traced back to the behaviorist conception of conditioning. In behaviorist versions, associations connect external stimuli with motor responses; both of these are external, physical events. The nervous system, then, is nothing more

82 MODELING

than a switchboard. The association razors anything we might now recognize as thought or information processing.[6]

While this claim is made more complicated as models have increased in complexity (as described in the last section), the idea that associative processes are (in some relevant sense) simple has arguably been definitional all along (Dacey 2019a). Associative models of animal cognition are common. The standard view is that pretty much all behaving animals have at least some simple associative processes, and some animals also have more sophisticated cognitive processes (in contrast, the assumption has arguably been that most human processes are cognitive, while some might be associative). Thus, animal minds are presumed to be simple because associative models of animal minds are common, and associative models are common in part because animal minds are presumed to be simple. The two thoughts are mutually supporting (e.g., Meketa 2014).

However, looking deeper, I argue that we can see that the reasons for the ubiquity of associative models need not have anything to do with the (purported) fact that associative processes are common. Firstly, the preference for simpler processes enshrined in Morgan's Canon means that associative explanations of behavior are, effectively, the default. Whenever an associative process can be applied, it is preferred, because associative processes are generally seen as the simplest on offer (e.g., Shettleworth 2010; see also Halina 2022). Secondly, the basic *formalism* of associative models is extremely flexible and adaptive. As such, it can be applied often. Putting these together, an associative model (of some sort or other) often can predict behavior, and whenever they can, the associative model is preferred.

The experimenter's challenge, then, is to discover a behavior that the current associative model doesn't predict. This is what gets the iterating process described in the last section off the ground: some behavior is discovered that is not predicted by an associative model, which is initially interpreted as evidence for causal reasoning, until a new associative model is constructed that subsumes the result, until a new effect is discovered that *that* model does not predict; and so forth. "Trophy hunting" is arguably an attempt to end-run this process by demonstrating a behavior no associative model could explain. This is the associative versus cognitive dynamic that has been increasingly decried as unproductive by philosophers and psychologists alike (Penn and Povinelli

---

[6] One might wonder why this idea has dominated so. The main reason is that the idea itself has only recently been challenged and alternatives articulated (Buckner 2011, 2017; Dacey 2016b, 2017b). This is itself largely a result of how deeply embedded the idea is. It is often implicit in the design and interpretation of experiments themselves.

2007b; Buckner 2011, 2017; Smith, Couchman, and Beran 2014; Hanus 2016; Dacey 2016b).

The key claim I'd like to highlight here is that the basic formalism of associative models can be a powerful descriptive tool, *even if it does not fully capture the essence of the process involved*. Put bluntly, nearly anything can be described with some architecture of links and nodes. This fact alone tells us very little about the world.

To get more specific, let's walk through the interpretation of associative models. We can separate two kinds of models. Associative 'thinking' models describe the sequences that occur in perception, thought, and action. Associative 'learning' models describe the impact that variables relating to patterns of sequence have on learning. A single model can include both kinds of claim (and perhaps the paradigm cases do), but even then the claims are distinct. The claim that associative models must describe a particular *type* of process implies that these descriptions capture the essence of the process described: associative thought is merely following a pre-established chain of representations, and associative learning is only sensitive to the variables of co-occurrence present in the model. But these are additional claims that are built onto the interpretation of the models and are in no way necessitated by the fact that they describe these sequences and/or the impact of these variables. We can drop them and say that associative models simply describe regularities in the sequence of representations in a process (with little commitment about the mechanisms driving the sequence), and associative learning models describe the impact a certain set of variables have on learning (with little commitment about what other variables might also make a difference).

If we take this seriously, we see that the associative model can be valuable in describing processes of any kind and of any complexity. There are many purposes for which we would want to describe only the sequence of representational states that a process moves through, or the way that certain kinds of variable impact learning. In such a role, associative models provide a highly abstract, partial description of what the process does. In the context of causal cognition, an association might stand in for a connection in the mind that *has* representational content of a causal relation. An associative learning model can include variables that might relate specifically to causal relations (relative timing, contact, and so forth). Crucially, this means that associative models do not exclude cognitive models. An associative model might describe the sequence of states a process moves through, while a model of causal reasoning provides a fuller articulation of how or why it moves through that sequence. On this view, whether we use an associative or cognitive model depends on the

84 MODELING

question we are asking, not the process we are describing. We can discard the idea that there is a particular *kind* of processing that is distinctly associative (and denoted by all proper uses of associative models). Adopting terminology from Dacey (2016b), I call my view the "abstract description view" in contrast with the "associative processing view" that has driven the dynamics described in the last section (for a summary of the candidate views of association discussed in the chapter, see Table 3.1 below).

On my view, associative models are partial descriptions of the process not just because they isolate a single aspect of a process and describe it alone. They are partial also in the sense that (as a result of this) they can be rightly applied to a process in a single task and context even if they *only* get predictions right in that task or context. The practical manifestation of the idea of associative processing is the assumption that a single associative model is only properly applied when it describes what the process does in all contexts in which it operates. The role of this assumption can be seen above; evidence of some causal reasoning ability beyond the scope of an associative model is taken to invalidate associative explanations of all the previous variants of the experimental task (e.g., forward blocking can invalidate RW as a model of previous results showing backward blocking). This is, in practical terms, what it means for associative processes to be simple and inflexible. Even as the actual demands of proper task performance changes, the process is incapable of breaking free of its fixed course of response.

My view implies that an individual associative model should only be expected to predict in the task and context for which it was built. It may generalize. Presumably, individual models will generalize at least somewhat to slight variations in task and context. It's even possible that one may generalize fairly broadly. But they should not be *assumed* to generalize in all proper uses. The important point here is that an associative model's failure to predict in a new variant of a task need not imply that it did not accurately describe what was going on in the original variant.

For flexible, complex processes, we may need several associative models of a single process that describe its performance across tasks and contexts. These would describe the process collectively, rather than relying on any one model. This is a significant change to the way associative models are generally viewed, but it is really just to accept that the models only provide partial descriptions.

Associative models, in this role, can do more than merely coexist with cognitive models; they can actively help the construction and testing of precise, predictive, even mathematical cognitive models. The way they do so is, in effect, to mediate between a cognitive model and the data. Working from data to model, the associative model would summarize data and set up constraints that need to be met by any candidate cognitive model (it must produce those sequences/learning patterns). Working from model to data, an associative model can be built to make some behavior of a cognitive model in some context more precise and generate specific testable hypotheses.

The overall approach here would be to collect sets of associative models of how the process behaves in various experimental contexts and attempt to determine what sort of mechanism could produce those sequences. If that underlying mechanism is some sort of cognitive process, then the two kinds of models coexist in describing the same process. In this project, subsequent failures to predict do bear on evaluation of the explanation a model played in previous experiments. However, it is not the one-shot falsification it is generally considered to be. Instead, the question is whether there is some possible model of the underlying mechanism that would generate all the associative models that do predict. In other cases, there might be several associative models that predict in any specific context. In these cases, the associative models can again be evaluated alongside the cognitive models that do, and do not, predict them.

This would all be part of a more holistic examination of the evidence as described in Chapter 1. In fact, this was exactly the process I described for chimpanzee mind-reading (see Figures 1.3 and 1.5). Those were gestural, imprecise associative models, and I did not describe them as such there, but they were associative models nonetheless.[7] I suggest this strategy may be helpful in mind-reading, causal reasoning, and perhaps across comparative psychology.

Stepping back, the key claim here is that the fact that behavior is predicted by a model that includes associations is not reason to conclude the process belongs to the class of associative processing. If we follow through with this thought, and recognize associative models to be partial, abstract descriptions, the problematic dichotomy between associative and cognitive models dissolves. More to the point for worries about the lack of specificity of cognitive models, associative models can lend their precision to the cognitive models. A model of a cognitive process combined with a suite of associative models will amount to a much more precise understanding of the process than a verbal characterization alone would allow. This strategy could also work to characterize the gap between association and cognition as typically conceived, where even verbally characterized models are lacking.

---

[7] If the approach seemed fruitful to you there, this chapter describes the way to make it precise and generalizable. If the approach did not seem fruitful there, hopefully this fuller articulation makes it more plausible.

86   MODELING

While the details of the view might be fairly complicated, my argument for it is simple. First, I point out that such an interpretation is possible. The interpretation of associative models need not include the claim that they denote associative processes. Second, I argue that the abstract description view that arises with this reinterpretation will likely be more fruitful than the associative processing view. This step of the argument is a bet about what will be most useful going forward. I think that this sort of argument is where any debate between different possible ways of interpreting a model will go. Nonetheless, there may be reasonable disagreement on this claim. Specifically, one might wish for models to provide greater explanatory depth than associative models do on my reading, a point which I turn to now.

3.5 Mechanistic Explanation and the Aims of Modeling

In the Introduction I framed the project that I am most interested in as being the discovery and understanding of the *mechanisms* responsible for animal cognition. 'Mechanism' is, at this point, something of a term of art in philosophy of science, and it will help articulate my understanding of associative models, and potential challenges to my view, if I explain a bit more about what mechanisms and mechanistic explanation are and what they entail.

Mechanistic explanation is a strategy that explains how some phenomenon is produced. It proceeds by, first, breaking the system that exhibits the phenomenon into parts and, then, showing how the activities performed by the parts produce that phenomenon (Craver 2007). Thus, mechanistic explanation involves a decomposition that is both structural (the parts one identifies had better be there) and functional (the parts you identify will depend on the activities that are relevant to the task at hand). In psychology, a decomposition may proceed largely functionally if the structural parts are not known well, but this will carry with it some commitment to the fact that the functions one identifies will be realized somewhere, somehow, in the system (Cummins 2000; Piccinini and Craver 2011). Neuroscience, in contrast, has much more direct access to the parts.

The easiest examples to understand generally involve machines: if we want to understand how an internal combustion engine generates torque, we look to the parts and what they do: the spark plug ignites gas in the chamber, which drives the piston down, cranking the driveshaft, which spins. So the parts are partially identified functionally; they are not simply cube-shaped bits of the engine, they are parts that perform identifiable functions. And, if it turns out that the parts are not there, this explanation is wrong. For example, diesel engines do not have spark plugs (the fuel ignites due to compression and temperature), so this would be the wrong explanation of how diesel engines work. However intuitive this example is, though, we should not take it to be comprehensive; mechanists are committed to the claim that these decompositions can be performed in the messy 'wetware' of a brain as well. The 'parts' in such a system may be harder to identify than in a manufactured machine, but they are there. In a brain or cognitive system, the relevant parts may be brain regions, networks of brain regions, neural circuits or networks, individual neurons, and even individual parts of the neurons themselves (ion channels, axons, synaptic terminals, etc.).

Note also that any part performing an activity can, itself, be treated as a phenomenon we might wish to explain (how is it that the spark plug sparks?). As a result, mechanistic explanation is characteristically multilevel. If we aim to understand some performance by a behaving animal, for example, we might identify some activity being performed by a part of the animal's brain. If we aim to understand how that part of the brain performs the activity, we might note activities in a particular neural circuit. In turn, we might explain those circuit phenomena by noting activities in individual cells, and so forth (as depicted in Figure 3.2). As a corollary, this means that mechanistic explanation is always interest-relative in the sense that the mechanism one identifies in any system will depend on which phenomenon one wishes to explain.

Taking associative models to describe a mechanism responsible for behavior means taking it to be the case that the representations are in the brain, somehow, and that they do follow the sequences specified. On the learning side, it means taking it to be the case that the system somehow actually encodes information about patterns of co-occurrence, and that encoded information has a causal impact on the sequences of representations the process follows.

My view takes the mechanistic commitments of the models to stop there. It doesn't say anything about the kinds of representations involved or the mechanisms by which they follow the sequences they do. The associative processing view, on the other hand, takes the models to carry much more significant mechanistic commitments. For example, one might think the associative link is brute causal; there is no structured mechanism underlying it except perhaps literal neural connections, as in a scaled-down version of the behaviorist neural switchboard.

88 MODELING

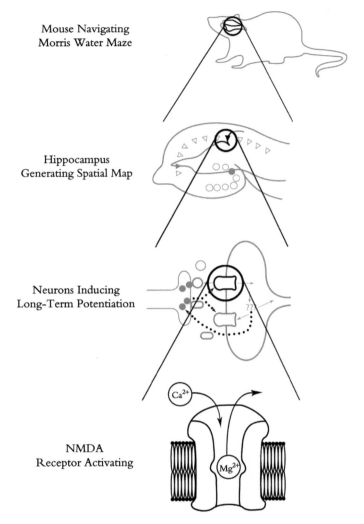

Fig. 3.2 Example of a multilevel mechanism for spatial memory. Reprint of fig. 5.1 in Explaining the Brain by Carl F. Craver, © Carl F. Craver 2007.

My view might start to look instrumentalist to some. It's not. It's realist in the sense that is most important here: it implies a commitment to the accuracy of the models regarding sequences of representations and the influence of the stated variables on learning. That is, it takes the sequences and learning effects to be *real*. The standard of evaluation for the model is not just behavioral prediction, as an instrumentalist position would hold. If it turns out that these elements are not present and are not doing what the model says, then the model is wrong.[8] Still, associative models are only partial, abstract descriptions of what is going on in the system.

For comparison, suppose that we were trying to understand how an internal combustion engine works, and the only instruments we have record the movement of air molecules. We might then produce a model that depicts air rushing into the combustion chamber, followed by a flurry of chaotic expanding activity, followed by air leaving the combustion chamber in a different way than it came in. This model would not tell us everything that is going on, far from it. However, we could be realists about the model; we could take the air to be real and really doing what the model says. If it got the air movements wrong, we would reject the model. Moreover, the partial model would set constraints on any more complete model of the engine's mechanism: that mechanism would need to move the air around in the way our model describes.

As above, the machine example is simpler than the cognitive system example, and the difference does matter. The project of associative modeling I have described involves generating candidate sequences of representational states based on behavioral observations and then working down to candidate mechanistic models of the cognitive processes responsible. This kind of top-down approach has been criticized by some as implying that psychology is *autonomous* from neuroscience; that psychology can and should operate with no input from neuroscience (Boone and Piccinini 2016). Autonomy dominated the philosophy of cognitive science and philosophy of mind for most of the twentieth century. It is now, however, widely seen as outdated.

Regarding the models more specifically, one might also object that we want models that come with greater mechanistic detail, because those models provide greater *explanatory power*. Explanatory power, here, involves the number of counterfactual situations one can predict and explain, the precision of the model, and the degree to which it integrates with other models and areas of background knowledge (Ylikoski and Kuorikoski 2010). By any of these measures, greater mechanistic detail will usually increase explanatory power.

And indeed, the abstract description view is not the only route available to us. Cameron Buckner and I, for example, agree broadly on the problems that arise as a result of the associative processing view, and about the importance of addressing them (Buckner 2011, 2013, 2017). We also agree that associative models can play an important part here. However, we go in opposite directions

---

[8] I don't intend any fancy version of realism here; I simply want to differentiate this view from a purely instrumentalist view that only evaluates models on whether they get predictions right, not on whether they make accurate claims about the internal workings of the system (I mention such an instrumentalist view as proposed by Colombo and colleagues in the conclusion of the chapter).

90 MODELING

in our proposed solutions. Instead of taking associative models to be more abstract, with fewer mechanistic commitments, Buckner suggests moving to more concrete "associative" models: those which describe sequences in actual neurons and neural circuits. This, he argues, also dissolves the associative/cognitive dichotomy because a sophisticated cognitive process can be *underlain* by 'associative' networks of neural activity (Buckner 2017). Indeed, there is a tradition of treating neural models in connectionism or in computational neuroscience as being the modern standard-bearers of associationism (Bechtel and Abrahamsen 1991; Clark 1993). This we might call a reductive or neural view of associationism (see Table 3.1).

The idea that neural models implement associations is based on a couple of key similarities. First, they share a general network structure. Second, it is known that connections between neurons are strengthened by repeated use (Hebbian learning) in a way that looks a lot like associative learning by co-occurrence. The key difference in the models, though, is that in associative models, as I use the term, the nodes in the network stand in for representations. In neural models, the nodes stand in for neurons or groups of neurons. As such, the models say different things about different parts of the system. I take the difference in what the model says to be more important than the formal similarities; after all, I suggested above that network formalism is an extremely flexible descriptive tool for lots of purposes.

So, there are two possible routes which, in principle, can accomplish the goal of dissolving the associative/cognitive divide and generating models of cognitive processes that are much more precise and concrete than we mostly

Table 3.1 Summary of the three views of association and associative models discussed here (see Dacey 2016b for more detailed comparisons).

View	Nodes stand in for	My assessment
Associative processing view	Representations (in a distinctly associative process)	Responsible for unproductive associative/cognitive divide; should be rejected
Abstract description view	Representations (in a sequence with no commitments to underlying mechanisms)	The most promising approach; suggested here
Reductive/neural associationism	Neurons or collections of neurons	A valuable set of models, but casting them as the new standard-bearers of associationism risks ignoring abstract description view

have today. The question is which approach is better. Here, again, I think the only actual answer is to make a bet on which approach is more likely to work. And I think that this represents a significant difference in viewpoint between me and Buckner; he is comparatively more optimistic about the current state of neuroscience, and I am comparatively more optimistic about the prospects of a somewhat more traditional form of psychology.

In defense of my own view, I will first say that I do absolutely encourage development of various kinds of computational neural models. I just think, *in addition*, the approach that makes use of associative models as abstract descriptions will be fruitful. My overall strategy is pluralistic about the methods and model types that should be pursued. The key is to recognize the differences between types of models and the differences in the kinds of claim they carry about the target system (in current practice, they are often run together, and I worry that a full embrace of neural associationism encourages that). For reasons having to do with the history, I have argued that 'abstract description' models retain the name 'associative,' while *connectionist* or *neural circuit* models be named as such (Dacey 2016b, 2019b).[9]

I'll note, as well, that my approach does not imply any pernicious sort of autonomy of psychology, because it is pluralistic. Ideally, we'd be able to describe a single process with many such models at the same time, each doing different work (e.g., Hochstein 2016). Some might be abstract associative models, some might be cognitive models, and some neural models; these may be ordered along the hierarchy of mechanistic levels, Marr's (1982) levels, or they may have other kinds of relations that set constraints (Danks 2014). In any case, they all set constraints on one another. At the same time, they all must be able to operate in some degree of isolation, as each brings something different to the table.

The psychological autonomy thesis that dominated the functionalism of the twentieth century was one in which the brain *truly did not matter at all* for the progress of psychological science: As Putnam (1975) said, "We could be made of Swiss cheese and it wouldn't matter." In rejecting this view, and anything like it, we should not slip into reductionism or neuroscience imperialism. There is still work for psychology to do.

More particularly, I think the actual information encoded by a process is crucial to understand what kind of cognitive mechanism it is; that is, a capacity counts as causal reasoning because it allows reasoning about causal relations

---

[9] I stand by these arguments, though I care more that the meaningful differences between models are respected than I do how the terminology is applied.

qua causal relations, and it is mind-reading because it allows representation of mental states in others *qua* mental states in others. In Buckner's (2014) own terms, these are problems of psychosemantics. Buckner (2013) argues that there are features of neural organization (that is, specific kinds of neural circuit) that reliably allow and indicate the kinds of flexibility that are usually attributed to cognitive processes as opposed to associative processes as that category is usually conceived. Even if this were granted, the circuit organization cannot tell us exactly *what* the cognitive process in question is doing. In order to answer that question, we need to carefully study exactly what information the system encodes and how it reacts to that information. Associative models, as I describe them, do just that. This allows a much finer grain of differentiating the content of the system than any straightforwardly neural approach can allow (and will for the immediate and medium terms; over time, progress may allow greater integration of these sciences than this picture describes).

However, I am also skeptical that we will end up with a clean division between types of processing, and I suspect a spectrum of complexity/flexibility is most likely (along at least two dimensions; Dacey 2019b). As such, I'd generally prefer that the category of 'associative processing' be dropped entirely, as I find it misleading: associative processes are typically defined as the processes denoted by associative models.[10] So if we drop the idea that the models denote a category of processing, we should drop the category that is defined by this supposed denotation. I generally prefer the term 'psychological processing' as a more capacious term for anything characterized in broadly psychological (information-processing) terms, and I see no reason why the term 'cognitive processing' can't be applied similarly.[11]

In any case, the pluralist approach here is one reason that this view of modeling need not lead to eliminativism about the mental, as might be worried. It is fine to retain verbal labels, like causal reasoning or mind-reading, while allowing *other* models to do the more precise work.[12] This approach requires collaboration, integration, and negotiation across areas, for various models to be ultimately brought into alignment. This project is difficult, but that's not its problem alone; this entire book is predicated on the difficulty of the science of

---

[10] As we see in the way that the concept of associative processes has been continually retrofitted to new models as they come up.

[11] It still makes sense to think of 'associative' models as a kind of model based on shared formalism, though this view makes the boundaries here a lot fuzzier: since associative models and cognitive models are not mutually exclusive, models can unproblematically include elements of each type. Indeed, I see it as a strength of the view that these mixed models fit more naturally.

[12] As, for example, we may verbally describe some process in cellular biology but also have pictures, visual 3-D models, pathway models, mathematical models, and chemical/physical models, all to scaffold that verbal description.

animal minds. The bottom line is that this pluralistic, multi-model, multilevel approach to modeling cognitive processes is most likely the only way to make progress.

3.6 Conclusion

The core claim of this chapter is that associative models have been wrongly interpreted as denoting a particular *kind* of psychological processing. Instead, they should be taken to provide only partial, abstract descriptions of the phenomena and/or mechanisms involved. This means that they are compatible with other kinds of mechanism. This strategy breaks down dichotomies that have been problematic, such as the associative/cognitive divide. It provides what is, in effect, a new tool for describing processes in the "gap" of moderate complexity, and for providing mathematically precise models of cognitive processes. It contributes to a pluralistic, multi-model strategy of understanding cognitive processes.

Reinterpreting associative models is not the only way to go. As noted in Chapter 1, dimensional models such as that proposed by Starzak and Gray (2020) provide another—one which can be pursued alongside the strategy described here. Insofar as the key idea is to interpret a prominent kind of model in a way that is more modest in its commitments, the strategy might also generalize. For example, Bayesian models are a class of models that describe possible ways of rationally drawing conclusions based on information that is uncertain, incomplete, or noisy. The impressive predictive success of Bayesian models, like associative models, has motivated claims that argue that the brain, in general, is a Bayesian machine (e.g., Knill and Pouget 2004; Frriston 2012; Clark 2015), setting up a dynamic much like that described here. Colombo and colleagues (Colombo and Seriès 2012; Colombo, Elkin, and Hartmann 2021) argue that the models should be interpreted instrumentally; they may get predictions right in many settings, but we should not interpret them as telling us what the brain is doing.

While the details are different,[13] there is a similar spirit here regarding a more modest interpretation of the mechanistic commitments carried by the models. Whether this is right, and whether this strategy of modest interpretation

---

[13] Crucially, their antirealist stance is different from my realist stance towards associative models. Part of their argument is skepticism that Bayesian models will turn out to be the most predictive models of uncertainty management (which I share for Bayesian models, but not for associative models).

applies to other models, are open questions. But it is worth looking into. This modest attitude towards models helps us use models in ways that make more specific predictions in more limited contexts, thus showing how individual pieces of evidence can be taken to do real work without actually settling the big questions. The result, I think, can be a more productive science overall.

4

Integration and Homology

4.1 Introduction

Cognitive science was formally announced as an interdisciplinary field of study in 1978 by the Sloan Committee meeting and report (Keyser, Miller, and Walker 1978; Miller 2003). The report describes six fields with varying degrees of interconnection: philosophy, psychology, neuroscience, computer science, anthropology, and linguistics (depicted in the famous Sloan Hexagram; Figure 4.1). This was aspirational as well as descriptive. The authors thought that the best route for growth was by building more connections between these fields, and perhaps welcoming more into the fold. In the years since, cognitive science has pursued that goal, although it is debatable how successfully. While some are optimistic that true integration is well underway (e.g., Boone and Piccinini 2016), others are less convinced (e.g., Nunez et al. 2019; see also responses in *Topics in Cognitive Science,* 11(4), 2019). Either way, the goal is broadly shared, and cognitive science continues to define itself as an interdisciplinary undertaking.

Research on animal cognition is a microcosm of cognitive science in this regard; comparative psychology, philosophy, neuroscience, ethology, and evolutionary biology have arguably been the primary contributors. Even so, at least in the philosophy of animal minds, there has been a growing trend to call for greater integration. These typically call for a greater role for neuroscience and evolutionary biology in comparative psychology (e.g., Dacey 2016a; Fitzpatrick 2008; Mikhalevich, Powell, and Logan 2017; Andrews 2020).[1] I agree with these calls. We should take evidence from wherever we can. When we recognize the limitations of any single piece of evidence (as I have argued for), new sources become more valuable.

Cross-disciplinary integration is always a challenging task. Different fields ask different questions about different parts or aspects of animals and their lives. They use different methods that produce different kinds of results and organize them into different kinds of models and theories. Minimally, though,

---

[1] Ethology has also been a popular suggestion, which I will address in the next chapter.

Seven Challenges for the Science of Animal Minds. Mike Dacey, Oxford University Press. © Mike Dacey 2025.
DOI: 10.1093/9780198928102.003.0005

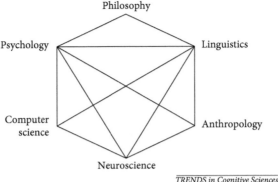

Fig. 4.1 The "Sloan hexagram" of interdisciplinary connections in cognitive science in 1978. Reprinted from George A Miller, "The cognitive revolution: a historical perspective," *Trends in Cognitive Sciences*, 7(3), 141–4, © 2003, with permission from Elsevier.

evidence from one field can set constraints on another, narrowing the space of hypotheses that should be seen as candidates, or providing some reason to think one candidate hypothesis is more plausible than another. More systematic frameworks might provide a common language for models across fields. New tools and methods can bring measurements from different fields closer together. There are as many ways that integration can go as there are possible integrations, so I cannot address all the ways this can be difficult. Instead, I focus on a specific challenge that I see as pervasive and underappreciated: the ways that integration of neuroscience and evolutionary biology into comparative psychology often require attributions of *homology*.

Inferences that integrate across fields often require cross-species comparison, using something that we know about a trait in one species to inform our understanding of a similar trait in another species. These inferences usually work best when the traits are *homologous*: Homologous traits are those descended from a single trait in the most recent common ancestor of the two species.[2] For example, human arms are homologous with bird wings because they both evolved from forelimbs in our most recent common ancestor, a reptile-like amniote, about 300 million years ago. Despite the differences

---

[2] I do not claim that every integrating inference is hostage to homology, nor do I take the discussion to exhaust the interesting things to be said about cross-disciplinary integration. I do take homology to be an issue of particular importance, especially since it is not sufficiently appreciated in this context.

between arms and wings, there are commonalities that these traits share because of their shared lineage. Perhaps the most obvious examples are commonalities in bone structure. This is why homology attributions can support claims that traits are likely similar; recognizing traits in different species as homologous means they are more likely to share as-yet unobserved features. This similarity is not guaranteed, and we need to be careful about the kinds of similarity we attribute (my arms' homology with bird wings sadly doesn't mean that I can fly). Even so, homology is a powerful and important source of evidence regarding traits we do not understand well: comparative inferences supported by homology attributions allow integration of evidence across target species and disciplines of study.

I'll discuss three specific examples of attempted integration which each require homology attributions in different ways. These include the use of neuroscience to understand emotions such as anger (Section 4.3), the use of evolutionary parsimony considerations to argue for social understanding in chimpanzees (Section 4.4), and the use of claims about problems likely faced in the evolutionary past to make inferences about how the mind works today (Section 4.5). The different ways these require homology demonstrate different problems integration can run into when the difficulty of homology attribution in psychology is not fully appreciated.

That *comparative* inferences are important for a field known as *comparative* psychology may not be surprising. However, the role of homology claims in these inferences, and thus in many (perhaps most) attempts to integrate work from neuroscience and evolutionary biology to comparative psychology, has not been appreciated. Too often homologies are simply assumed, or the role of homology in an argument is ignored. Even when we have good reason to think some homology holds, it can be unclear what the exact homology is, and thus what we can conclude from it. It is especially hard to identify and characterize homologies when we are dealing with cognitive or functional homologies, as opposed to structural, especially skeletal, homologies which can be traced through the fossil record. Recognizing cognitive homologies requires outlining the cognitive capacities in each species (a difficult enough task, as we've seen), and then mapping between them despite the near certainty of an imperfect comparison (and perhaps even harder, mapping back to a common ancestor).

Thus, the challenge of cross-disciplinary integration requires addressing the challenge of homology attribution, and so I consider them together in this chapter. The challenge of homology, in turn, manifests in two specific problems. First, a *boundary problem* of tracing the boundaries between one candidate homologue and another, and thus to identify which homologue is the right target. Second, a *specificity trade-off* in which making homology claims either more specific or more general trades one inferential weakness for another.

4.2 Homology Basics

Homology has been a subject of increasing interest in philosophy in recent decades, for several reasons. One is the idea that 'homology thinking' represents a unique mode of biological thought (Ereshefsky, 2012; Wagner 2016), and a central organizing principle in biology (Currie 2021). Homology also provides a window into core questions about the nature of the evolutionary process, including the way selection acts on traits we characterize functionally, structurally, or developmentally (Brigandt 2003; Ereshefsky 2009; Wagner 2014), and the relative degree of constraint we take to come from the environment or from the organism itself (Powell 2007). There have also been prominent calls that psychological categories, such as the emotions, should be treated as historical kinds that are defined by homology rather than by functional role (Griffiths 1997; Matthen 2007; Ereshefsky 2007). While my arguments fit in the broader context in several ways, my main interests are more focused. My goal here is simply to show that homology assumptions underly many (perhaps most) inferences that are used to integrate neuroscience and/or evolutionary theory with comparative psychology. Given that this is the case, we need to respect the difficulty of identifying cognitive homologies when we make these integrating inferences. Some basics of homology—what it is, how to identify it, and what can be tricky—will be helpful before we discuss actual attempts at integration.

Two traits in members of different species are homologous when each is derived from a single trait in the most recent common ancestor. The easiest examples to understand are probably skeletal structures. Basic skeletal structures of vertebrates, such as arms and wings, share similarities that can be made clear with a trip to the natural history museum (this is why those trips are such good, if perhaps macabre, ways to make yourself feel similar to other animals). Sometimes traits that evolved independently are similar because of shared selective pressures. These traits are called *analogous* rather than homologous. For example, fish and whales share a 'torpedo' shape, even though whales have land-mammal descendants that looked like hoofed wolves, while fish ancestors never left the water. As such, evidence of homology is not the *only* way that evolutionary considerations can indicate that traits are similar. However, homology attributions typically occupy a special role for present purposes because they are called upon when more direct evidence is indeterminate, making comparisons with related species valuable.[3]

The core question, for our purposes, is when and how we can make inferences from what we know about members of one species to members of another. As an example of the kind of inference I have in mind, imagine we discover a new mammal species that uses its front limbs in an unusual way. Perhaps we are encountering moles for the first time, which hold their front limbs with their paws facing laterally away from their body and dig by pushing their arms forward into the earth and out. We might identify these limbs as homologous with other mammal forelimbs and on that basis make reasonably confident, though defeasible, predictions about the number and orientation of bones in the limb. As long as this kind of inference is justified, my focus here allows me to sidestep some of the thornier issues relating to homology, such as those about its ultimate nature. Structural homologies, such as forelimbs, enjoy less central status today than in the past, and some think that *real* homologies ought to be identified in genetic/developmental mechanisms (Rosenberg 2006). The idea of functional, behavioral, or cognitive homology is especially controversial (Love 2007; Ereshefsky 2007; Wenzel 1992). I adopt an expansive meaning of the term for present purposes and will be satisfied as long as a homology attribution can support the inference that the two traits are likely similar in as-yet unobserved ways.[4]

There are some ways in which cognitive homologies are especially difficult. For a start, it can be hard to identify homologies in the first place. While there is no generally accepted standard for identifying homology, Remane's (1952) list of three criteria provides a common starting point. The first is the *position* of the trait. For instance, the humerus bone (upper arm) retains its relative position to wrist bones across species. Second is *special quality*. The idea here is that traits that share specific features are more likely to be homologous, especially if the shared features are incidental or counterproductive to the function served. For example, some wrist bones in humans don't serve any real purpose.[5] As an example of something genuinely counterproductive, the blood

---

[3] In cases where analogy is appealed to towards similar ends, the work is usually being done by appeals to adaptive constraints rather than any direct comparison of the traits. Of course, this need not always be the case, and my interest in homology is motivated not by the idea that it is the only tool for comparison, but by the significance I see in the way it operates in the field.

[4] Readers with specific commitments here can take this as a simplification of language: perhaps 'proper' homologies would be better in many ways. They might even support stronger versions of this inference, but the discussions here address functional, behavioral, and cognitive traits, so those are the comparisons we are considering (see also note 6).

[5] The hamate bone in the wrist is prone to fracture by baseball players and golfers. When this happens, it is often simply removed with no long-term repercussions.

supply in the vertebrate eye occludes part of the retina; that this organization is so broadly shared is evidence of homology. Third is *continuity of intermediates*. For example, human and bird forelimbs look quite different, but we can identify intermediate versions that bridge these differences, both by surveying the fossil record and comparing living species.

Applying these criteria to cognitive homology might require some changes. For example, *position* need not be physical. The basic idea can reasonably be applied to functions, such as behaviors and information-processing capacities, that occur in specific positions in a sequence or may be associated with or dissociated from one another in characteristic ways. Love (2007) suggests that a more abstract version of this criterion, applied to psychology, would be *organization*. For example, Wenzel (1992) notes that tail movements in two species of tilapia look different but occupy the same place in dances, so are considered homologous. One could also liberalize *special quality* to include functional features, not just structural ones. There may be quirks to the way some process operates (for instances, shared biases across species might be counted as a shared special quality). Continuity of psychological intermediates might be identified across living species if we know enough about a broad enough range of animals. Given the state of knowledge, this continuity is likely to be vague at best, for now.

When identifying homologous traits, we also need to set 'boundaries' in at least two different senses. As an example of the first boundary, we might be interested only in similarities in the hands of vertebrates, or we might be interested in the entire front limb. We might focus on the stomach alone or the whole digestive tract. Each of these might be perfectly valid homologues in a given group of animals, but they will give different evolutionary stories and might highlight different similarities and differences across species. I'll call this the 'individual spread' of a trait (the spread of the trait within the individual), noting that there might be multiple 'grains' at which we can identify a trait of interest for different purposes.[6] For cognitive traits, we might again replace the 'position-based' individual spread of traits with a spread of different functions (as we'll see below). The second boundary is the size of the group of species (the 'clade') which we take the homology to apply to. This is fixed by

---

[6] Wagner (2014) argues that the *actual* homology must be identified with a trait that is produced by a particular gene regulatory network ('character identity network'). Presumably this would fix the actual individual spread of a given trait if we knew the relevant genetic/developmental details. Even if we adopt this view or one like it, an imperfectly grasped homology can do the evidential work needed for the kind of cross-species inference that I focus on here: that some degree of similarity is more likely than if there were no homology. (And if one denies this, then my main point in the chapter becomes even more pressing, as it becomes substantially harder to identify the 'actual' homologies.)

how far back we take the most recent common ancestor to be. So, for example, if we are interested in the hand, we might be interested in the ape forelimb, including humans, chimpanzees, bonobos, gorillas, orangutans, and gibbons. Alternatively, we might zoom out a bit and compare forelimbs across a larger group; choosing the group of mammals would include all the apes mentioned, but also the limbs of bats, horses, dolphins, and so forth. I'll call this the *cladistic spread* of the trait, noting again that there are multiple different grains at which we can identify a trait of interest for different purposes. In the following, I'll simply refer to 'grain' or 'spread' when I intended it to be inclusive or non-committal about which of these two dimensions I mean, and indicate when I mean either *individual* or *cladistic* specifically.

These different kinds of spread are in principle dissociable, but for practical purposes, depending on the question we are asking, each might inform the other. That is, if we are interested in 'hands' as a functional category, it may be confusing to cast too broad a cladistic net and include horses and dolphins. But that diversity might become informative if we are interested in complete forelimbs. As a general matter, it's likely that we will find more similarity in finer-grain traits on both dimensions, and thus more specificity in the claims we can make: we can infer more specifically about human hands as they are now by looking at ape hands than we can by looking at mammal forelimbs.

As should be clear, the entire project of identifying homologies requires individuating 'traits' to compare across species. Things here are, again, harder in psychology: the best way to individuate psychological capacities, and understand their relations, is a difficult problem. One might hope that our psychological capacities are relatively sharply individuated, in the same way that the ends of specific bones set practical boundaries for various ways of individuating traits like arms and hands. We might also hope their functions are clearly distinct such that we can identify clear evolutionary steps that make a capacity possible, as perhaps the evolution of distinct digits, or opposable thumbs, mark important stages in the evolution of primate hands. Indeed, assumptions that a cognitive system is 'modular' in some significant sense are common in work that engages cognitive homology. Modular systems are relatively isolated from one another functionally and anatomically and are typically thought to be subject to significant evolutionary/genetic constraint. And as we will see in the three examples below, things are usually easier if the system actually is modular: at the extreme these systems might seem like Lego blocks that evolution copies (perhaps with small changes) and plugs into different species. Things are much harder if they are not so distinct. Plausibly, cognitive

102 INTEGRATION AND HOMOLOGY

systems can overlap functionally and in brain networks; they can be applied in novel ways, stretched to new tasks, and the brain can rewire itself continually. Across evolutionary time, differences are likely incremental as well; we cannot expect to find a distinct point at which a capacity came into being, all or nothing. So, it's hard enough to identify the boundaries of a cognitive capacity even in a single species. Comparing across species, in which the capacities differ in any number of ways, is even harder. Likely, we cannot simply read off the boundary from observations. We may have to make judgment calls about what counts as the capacity we have in mind, and what cannot. There may not always be a fact of the matter either, as different boundaries may serve different purposes. The space of options here remains wide open.[7]

But even in the most straightforward modular case, we do need to discover these boundaries and make sure our homology attributions respect them. This can be hard, and we cannot simply assume we have accomplished it based on superficial similarities. I'll call the general difficulty of identifying the right grain, in either the individual or phylogenetic sense, the *boundary problem* for cognitive homologies (borrowing Wiegman's term; Section 4.3).

In the next three sections, I'll consider my examples of attempted integration. Each requires homology in a different way and so helps us draw different lessons about how it can be difficult and what's at stake if we get it wrong. These worries generalize, but the current examples illustrate them well.

4.3 Neuroscience of Emotions and the Boundary Problem

Emotions are often thought to represent a particularly good target for which neuroscientific and evolutionary work can integrate with psychology. This idea comes mostly from a tradition of work on what are known as the *basic emotions* in humans. The standard list of basic emotions is happiness, sadness, fear, disgust, anger, and surprise. The idea is that these emotions are 1) evolved (more precisely, tightly genetically controlled), 2) relatively hardwired into our nervous systems, and 3) shared broadly with other species. Basic emotions are also often thought of as modular and cleanly dissociable from one another.

---

[7] I hope to remain noncommittal on where specific cases fall in this range from modular to muddled, but I recognize that even this noncommittal stance amounts to questioning assumptions of modularity. To be clear, despite this expectation of messiness, I do take cognitive abilities to be evolved in some nontrivial sense. My goal here is not to push for an extreme empiricism or reductionism; it is simply to lay bare the assumptions behind many integrations across fields.

These claims are clearest in the view that basic emotions are *affect programs*—automatic patterns of physiological activation. Each most famously is also taken to include a characteristic facial expression (Eckman and Oster 1979). I'll use anger as the example. Once the anger circuit is activated, a suite of automatic neurophysiological reactions follows: muscles tense, the heart rate increases, pupils dilate, the skin starts to sweat, and the eyebrows furrow into the classic expression of anger. None of this is controlled or conscious. And all of this makes straightforward evolutionary sense: tensing muscles and increased heart rate prepare us for immediate action, such as fighting; the facial expression signals our state to others, in the case of anger perhaps a warning that might help avoid a fight. This pattern of reactions overall could serve several adaptive roles in preparing and motivating behaviors that serve functions such as defense against attack, defending one's territory or access to mates, seizing another's territory or access to mates, enforcing one's position in a social hierarchy, or even enforcing the social rules of a group through punishment.

This approach to the basic emotions has been extremely influential, and perhaps even dominant, in the decades since it emerged. It is not uncontroversial, though. Some authors, such as Barrett (2017), argue that emotion categories are cultural constructions and/or learned interpretations of body states. I focus on affect program views because these are the best candidates for strong integration between psychology, neuroscience, and evolutionary theory. I won't argue either way, and my emphasis on affect programs shouldn't be read as agreement necessarily: indeed, constructivists may find my arguments amenable since they raise challenges to prominent affect program views.[8] One can also adopt an affect program view of basic emotions without applying it to all emotions. For example, Griffiths (1997) distinguishes basic emotions from 'higher cognitive' emotions, arguing that they are fundamentally different and that basic emotions in humans are homologous with emotions in other animals, but higher cognitive emotions are not.

Much influential work on the neuroscience of emotions has been done in model species such as rats and cats. There are good reasons for this. For instance, techniques for measuring human brain activity are typically very limited in spatial and/or temporal resolution. Ethical standards prevent interventions on humans such as lesions to brain areas, electrical stimulation,

---

[8] One might argue that the failures of homology in the emotion literature are a result of the fact that the affect program view is fundamentally mistaken; thus, it is a local failing and not a global challenge to the identification of cognitive homologies. But even so, the core worry remains—that this failure is, in large part, a product of underexamined assumptions of homology.

or single-unit recordings with electrodes in the brain. Even setting aside human-specific issues, research in neuroscience is time-consuming and expensive. As such it is difficult to motivate and fund work on lots of different species, instead of studying one model species and extrapolating. Methods are improving, but in the meantime work with model species continues to play an essential role. For our purposes, this means that any inferences drawn about the neuroscience of emotions in species other than the model species will require homology attributions. Most often, these inferences are used to make claims about the neuroscience of human emotion. However, the principle is the same for any species besides the model species. For instance, Panksepp takes his work on the neuroscience of emotions in cats to generalize to all mammals, and this assumption has grounded an expansive and ongoing research program (1998, 51–3; Panksepp and Zellner 2004; Siegel 2004; Panksepp and Biven 2012).

Panksepp's work is among the most ambitious and influential in the neuroscience of emotion, so it makes a good case study. His original favored list of "blue ribbon, grade A emotion systems" is somewhat different from the standard list of basic emotions, though his work is at the core of affect program views of emotion. His list includes SEEKING, FEAR, PANIC, and RAGE, and he later adds CARE, LUST, and PLAY (Panksepp 1998, 2011). We'll focus on the RAGE system, which produces anger. He identifies three brain areas as primarily responsible. These are the periaqueductal gray (PAG; part of the midbrain), the hypothalamus (part of the diencephalon, between the midbrain and the cortex), and the amygdala (on the inside edge of the cortex). Electrical stimulation of the right parts of any of these areas in cats will cause what is known as 'affective attack' or 'defensive rage,' which involves hissing, growling, and raised back hairs. This attack is generally thought to occur in response to threats. It is also distinct from the 'quiet attack' of predation, which is evoked by stimulation of different parts of the hypothalamus and PAG. I'll call this the *predation system* in contrast to RAGE. So the two kinds of aggression in cats seem behaviorally and neuroanatomically distinct, with affective attack a better candidate for anger. Panksepp also notes that activity generally flows from amygdala to hypothalamus to PAG, which then initiates motor outputs (ignoring some feedback connections). The amygdala takes inputs from the cortex, indicating that more elaborate 'cognitive' thoughts might activate the RAGE system. However, it need not initiate this way. Lesions to amygdala and hypothalamus do not suppress affective attack, but lesions to PAG can. So more basic physiological 'irritations' such as hunger can cause affective attack as well. This framework has inspired claims about emotions across the mammals, including work on how to treat and control aggression in humans (e.g., Siegel and Victoroff 2009; Blair 2012).

Wiegman (2016) introduces "the boundary problem" for cognitive homologies in response to Panksepp's view. He sets up the problem by contrasting the view of anger just described with another view, which makes a different, arguably competing, homology claim, coming primarily from behavioral experiments and ecological observation of rats.

The key behavioral differences that ground this homology claim are observed in territorial interactions between a resident rat and an intruder. Both behaviors in this case are more constrained and stereotyped than affective attack (though not completely). The first system is an *offense system* which drives behavior of the resident rat consisting in repeated attempts to bite the intruder rat on the back. The second is the *defense system* which drives counterattacks by the intruder rat, mostly aimed at the snout of the resident rat. Snout attacks seem to help prevent back-biting, create opportunities for escape, and they are paired with other behaviors that subserve defense from predators, such as freezing. Blanchard and Blanchard (1984, 1988, 2003) identify the *offense system* in rat confrontation as homologous to human anger, and the defense system as homologous to fear.

So we have two candidate hypotheses about homologues for human anger: one identifies the RAGE system in cats (opposed to the predation system), and one the offense system in rats (opposed to the defense system). Each of these hypotheses has something to say for it. The behaviors themselves seem to share features with anger—open aggression with seemingly destructive ends. In favor of the offense system, the context in which rat confrontation behavior is observed seems appropriate (or "fitting") to anger. It seems easy to imagine anger as a response to intruders into one's territory, and the response itself can defend territory and perhaps deter future incursions. In favor of the RAGE system, the same brain areas identified in the cat RAGE system, especially the hypothalamus and amygdala, seem to be involved in anger-like responses across mammalian species, including humans.

Unfortunately, the two hypotheses are likely incompatible. The RAGE system in cats and the offense system in rats are different in ways that make it unlikely that both could be homologues of human anger. In order for both to be homologues of human anger, they would need to be homologues of one another. There are behavioral and anatomical distinctions that make this seem unlikely. Behaviorally, offensive confrontation in rats is more focused than affective aggression in cats. While rats are somewhat flexible in how they accomplish the goal, offensive confrontation is aimed (specifically) at biting the backs of intruders that are (specifically) uncastrated males, and not killing attacks such as neck bites. Anatomically, it seems that the neural areas responsible for rat offensive aggression are slightly different from the areas responsible for affective aggression in cats (Haller 2013). In rats, the systems responsible for offense and defense involving conspecifics may not be cleanly separable (Siegel 2004, ch. 1), but there may be a distinction between systems responsible for aggressive behaviors aimed towards conspecifics in mating-related settings (that is, both sides of the offense/defense distinction) and those responsible for reactions to threats from predators (Canteras 2002).

These differences suggest that the RAGE system in cats and the offense system in rats are not (close) homologues to one another. It can be hard to know exactly what to conclude from this kind of evidence; which differences matter for homology assessment, and just how different do they need to be? Still, I follow Wiegman (2016) to the conclusion that the case against homology is stronger than the case for it. If this is right, it does not seem that both can simultaneously be (close) homologues to human anger. They are instead competing claims for homology to anger (see Figure 4.2).

For his part, Wiegman argues that Panksepp's hypothesis is built on a conflation of anger with fear. In humans, both anger and fear run through the hypothalamus. Wiegman argues that the specific *part* of the cat hypothalamus implicated in the RAGE system maps more closely with the part of the human hypothalamus involved in fear than in anger (Wiegman 2016, 230–3). So, the RAGE system might be a real and important part of the neuroscience of cat emotions, but Wiegman argues that Panksepp has identified the wrong homology. This is where the boundary problem comes in. It is difficult to tell

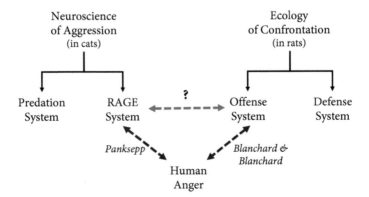

Fig. 4.2 Competing hypotheses regarding the homology of anger, based on Wiegman (2016).

whether the RAGE system in cats is homologous to anger, fear, or perhaps a coarser-grained, evolutionarily older (larger cladistic spread) system that evolved into both anger and fear. This coarser-grained emotion might be some kind of more general 'aggressive emotional arousal' (indeed, the hypothalamus and amygdala seem to be involved in emotional arousal with many emotions).

I take the status of the specific homology claim to be unsettled; the question of how to adjudicate these competing homology claims has not received much attention. Work by Panksepp and colleagues has been much more influential than Blanchard and Blanchard. While the Panksepp project has faced considerable criticism, that criticism typically attacks the idea of affect programs in general, not Panksepp's specific homology claims (e.g., Lindquist et al. 2012; Barrett 2017). My goal here is not to argue in favor of a specific interpretation but to show how the boundary problem arises, and how it ought to increase uncertainty regarding any comparative inferences here.

Even if we grant a modular view of fear and anger, there are three plausible homology claims at play: anger, fear, and a coarser-grained homologue including both. And if we drop, or loosen, the modularity assumption, things get messier still. Human emotions such as anger and fear may overlap and may have fuzzy boundaries even within individuals. These possibilities all impact not only whether a homology holds but also which homology holds, and, as such, what we can infer from the homology attribution.

In any case, actually demonstrating emotional homologies is a very difficult business, and we cannot assume that they hold based on immediately apparent similarities. What we need is a more systematic search for the criteria of homology: special quality, continuity of intermediates, and position/organization.[9] Wiegman proposes an additional constraint to address the possibility of a coarser-grained emotion being mistaken for a more specific one. He calls this the "class-specificity constraint":

> To provide evidence for relations of homology relative to homology class G as opposed to the more inclusive class, H, requires that some similarities between *relata* are not shared by traits in the more inclusive class, H. (Wiegman 2016, 235)

The way this plays out in any given case will presumably vary. For now, my emphasis is on the general shape of the problem.

---

[9] Panksepp himself (1998, 17–18) laments how difficult it is to get funding for work that would actually produce evidence for homologies, a call that I will echo below (Section 4.6).

Attempts to integrate neuroscience into understanding of psychological concepts often require substantive homology attributions because they require the use of model species. These homologies cannot simply be assumed, because it makes a difference which we take to hold, and it can be very hard to tell which does. This problem is not merely about needing data; we will need to decide where to draw the boundaries in our categories. Specifically, Panksepp and his followers attempt to integrate neuroscience with psychology because they use work on neural systems implicated in emotional responses to make claims about psychological emotional categories like anger. This requires homology because it requires the use of model species, from which conclusions are generalized to other species (in this case, mammals). Because any such homology attributions face the boundary problem, we should substantially increase our uncertainty about these conclusions until homology can be more convincingly established.[10]

It is worth noting that the need for homology claims is not specific to the integrative ambitions of this research program, but those ambitions substantially increase the challenge posed by the boundary problem. For instance, we could treat the RAGE system as a particular neural circuit, making no claims about connections to emotional categories. If we did, then the claim that other mammals have that circuit requires a homology attribution. However, the additional claim that other mammals have an emotion homologous to human anger produced by the RAGE circuit faces a much more difficult version of the boundary problem for two reasons. First, psychological categories are often much mushier than specific neural circuits, with more nebulous boundaries even in individuals. Second, the shift in levels between neural and cognitive claims opens up more opportunity for variation between species: the same circuit can serve a different function, and the same function can be produced by a different circuit.

Basic emotions are supposed to be among the best candidates for this kind of integration. And if things are this tricky even with basic emotions, imagine how much harder the issues are for more complicated reasoning capacities. These problems can get much more nebulous, and thus arguably harder, in attempts to integrate evolutionary biology with comparative psychology.

---

[10] Concerns like this may be raised against lots of work in biology that relies on model species, especially in biomedical research (e.g., Wall and Shani 2008).

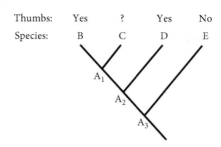

Fig. 4.3 A toy phylogenetic tree. Species B, C, D, and E are current; A1, A2, and A3 are ancestors. In this case, we have observed thumbs in species B and D, and observed their absence in E. Species C is as yet unobserved.

4.4 Cladistic Parsimony in Evolutionary Reasoning

The second common target for integrative hopes is evolutionary biology. There are a number of ways that evolutionary theory can impact thinking in psychology. Many of these involve some explicit comparison between species, and as such they require homology attributions.[11] I'll focus in this section on the use of *cladistic parsimony* to argue for claims about animal cognition.

Cladistic parsimony states that evolutionary changes which are adaptive, and thus retained, are rare. As such, we should posit them as sparingly as possible (Sober 1991). For our purposes, this can help make predictions about distributions of a trait across species. Specifically, the distribution of traits that requires the fewest evolutionary changes is most likely. There are some relatively straightforward applications that inform thinking about distributions of traits in current living species. If we know some trait is present in multiple close relatives of a target species, arranged in the right way on the phylogenetic tree, we can conclude that the target species likely has the trait as well. So, for example, in Figure 4.3, suppose we identify hands with opposable thumbs in species B and D; we can infer it is likely present in species C. The hypothesis that it is present in C only requires one change (between A_3 and A_2, while the hypothesis that it is not would require that the trait evolved twice, once on line A_1–B and once on line A_2–D). If adaptive changes are rare, the pattern requiring only one evolutionary change is more likely.

---

[11] Not all do. Many adaptive stories will not require homology, but these typically require other substantive assumptions about the constraints set by the environment and how those must be addressed. More minimally, I have argued that many parsimony claims ought to be read as arguments that evolutionary theory can be evidence for the plausibility of one candidate model of cognition (Chapter 1, Dacey 2016a).

110   INTEGRATION AND HOMOLOGY

These inferences are defeasible, of course, and should only be taken to provide some probabilistic evidence in favor of a hypothesis. Moreover, it should be clear how they depend on homology; the claim in the example would be that the trait identified in species B and D is a homology (originally possessed by ancestor A2), and as such it is likely present as a homologue in C as well.

Sober, who brought this mode of thinking into philosophical awareness, also provides a somewhat more ambitious version of the argument (2012, 2015). With it, he argues that parsimony favors the claim that chimpanzees can mind-read. That is, they can attribute mental states to others rather than simply tracking body position and orientation to predict behavior (see Chapter 1 for more). This version of the cladistic parsimony argument is based on the claim that *the same behavior* observed in members of two closely related species is evidence for *the same mechanism*. Sober adapts this argument from de Waal (1991), but he is clear that he thinks it is weaker than de Waal suggests. He does not think the argument is decisive, or even that it makes it 'likely' that chimpanzees mind-read. He simply takes it to be *evidence* in the rather expansive sense that it raises the probability that the hypothesis is true (an understanding of evidence that I share). This particular cladistic parsimony argument, I argue, runs into problems because of a particular instantiation of the boundary problem.

The basic idea behind this argument works. For a concrete example: if we know that humans carry oxygen through the blood (the function/behavior) using hemoglobin (the mechanism), and we know that chimpanzees must carry oxygen through their blood, we should expect that chimpanzees use hemoglobin as well. That would only require one evolutionary change sometime prior to our last common ancestor, of hemoglobin as a carrier of oxygen. If chimpanzees had evolved some other mechanism[12], it would have required two changes after our last common ancestor, one in each lineage, each producing a different mechanism for oxygen-carrying. Thus, cladistic parsimony drives the argument. Moreover, the argument is comparative at its core, requiring that a homology holds in order to make inferences about one species based on what we know about another related species; and the key is that the 'same behavior' argument provides evidence of homology, which in turn suggests a homologous mechanism.

In the context of chimpanzee mind-reading, the claim is that observations of 'the same' behavior between humans and chimpanzees in mind-reading

---

[12] There are other options; horseshoe crabs use hemocyanin, which is copper-based, making their blood blue while our blood is red from iron in hemoglobin.

contexts are evidence of the same mechanism. Since we know humans can attribute mental states, this amounts to evidence that chimpanzees can as well.

My worry with this argument is that it is not at all clear what it means to say that the behavior is 'the same.'[13] Sober never articulates the standards or reasoning that would be required to make this claim. It causes trouble because it might be impossible to articulate the claim in a way that is both plausibly true and informative about a possible homology relation. I'll try four articulations that I see as candidates. The argument doesn't fare well on any of them.

The strongest articulation would be that chimpanzee behavior is *indistinguishable from* human behavior. But even if we take chimpanzee performance on mind-reading tasks to be impressively sophisticated, this claim is a non-starter. Even setting aside questions about cognitive mechanisms, the social dynamics are different across species, and thus the behaviors differ.[14] On the opposite end of the spectrum, the weakest articulation would take the behaviors to be 'the same' merely in the sense that they are behaviors elicited by a task requiring social coordination or competition. This is too loose, as it wouldn't even require that we run the experiments to see what reaction we get; whatever response we see is a reaction to the task. Thus, the strongest articulation fails because it is unlikely that it actually applies across species; the weakest articulation fails because it applies too easily and doesn't tell us anything.

The third option would be to argue that the standards are based in theoretical understanding of the situation and the performance of the animals. This seems reasonable enough as a general thought. However, for this specific argument, theoretical understanding renders the comparative argument either irrelevant or circular. Suppose, for example, we say that both human and chimpanzee behaviors are 'the same' in the sense that they are both examples of 'mind-reading behaviors'—those behaviors that are responsive to mental states of others. However, if we set standards for this claim and observe behavior meeting that standard in chimpanzees, why not read that as direct experimental evidence of chimpanzee mind-reading? Moreover, if we try to

---

[13] As Sober puts it, "humans and chimpanzees both have behavioral trait B" (2012, 229). At times he seems to assert this, but at others it looks like it might only be intended as the antecedent of a conditional. Either way, though, the arguments to follow matter for any attempts to actually apply the argument form he is advancing. De Waal, for his part, frames it: "if their behavior resembles human behavior the psychological and mental processes involved are probably similar too" (1991, 316). So the same worry applies, though the fact that both sides of the conditional are framed as 'similarity' claims might imply more of a graded approach than Sober's framing does.

[14] Imagine, for example, that you were placed in the Hare et al. (2001) competitive feeding experiment in Chapter 1, with your boss on the other side, and cupcakes in the central arena. Do you immediately dive for the hidden cupcake and eat it so your boss can't take it first (as do chimpanzees)? Or do you greet your boss and comment on the weather before figuring out how to divvy up cupcakes?

112 INTEGRATION AND HOMOLOGY

make the comparative version of the argument, it is circular: if we describe the behavior in a way that presupposes the mechanism, we cannot take that description as evidence for homology and thus for shared mechanisms.[15]

The fourth articulation is a variant of the third. We could start with a theoretical understanding of mind-reading and what behaviors would provide evidence of it. But then instead of looking for exactly those behaviors, we relax the standards for members of species that are closely related to another species that we know can mind-read. That is, when chimpanzees display behaviors that look similar to mind-reading behaviors but still admit of ambiguous interpretation (as is the case in that literature; see Chapter 1), we should be more charitable in interpreting chimpanzee behaviors than we might in another, more distantly related, species.[16] From a homology perspective, this is probably the best version of the argument: it is true that similar traits in closely related species are more likely to be homologues than similar traits in distantly related species.[17] And, in turn, this may make other, as yet unobserved, similarities in the traits of closely related species more likely too. But even here, the argument runs into a significant version of the boundary problem: the plausibility of this argument, and what we take it to actually tell us, depend on the actual homology we take to hold and the types of evidence we have that it does. Because Sober simply asserts that the behavior is the same, he does not give us resources to answer either.

One could attempt to backfill some of this information to put the argument on firmer ground. As a general matter, though, we cannot just point to the empirical literature on chimpanzee social behavior, because that literature hasn't actually aimed at producing evidence for homology (and if we rely too heavily on theory, we run into the circularity worry for the third reading of the argument). But there might be some specific claims that help. For instance, it is

---

[15] This is a version of the general problem of behavior descriptions that presuppose intention: we might say I *reach for* my glass of water, or *glance at* my watch. We cannot take those descriptions, which presuppose the intention, as evidence of the intention. When Sober presents the cladistic parsimony argument in his book (2015, chs. 3–4), he raises the same issues about the role of mental states in grouping behaviors as a response to Dennett and behaviorism. He does not connect this discussion back to the cladistic parsimony argument, though, so he does not seem concerned about this.

[16] In the same discussions, Sober (2012, 2015) presents a different argument: that the mere fact of relatedness provides evidence that members of different species share traits. This argument is based on probabilistic causal reasoning: knowing that humans can mind-read increases the probability that our most recent common ancestor could mind-read, which in turn increases the probability that chimpanzees are capable of mind-reading. He points out that this requires no claims about chimpanzees themselves. Note a structural difference here: this is the argument that evolutionary proximity *itself* provides evidence of similarity, while the cladistic parsimony argument is that evolutionary proximity *changes the way we should evaluate behavioral evidence* of similarity.

[17] It is also probably the closest to what de Waal (1991) is doing in the passages Sober cites as originating the argument.

common in the mind-reading literature to view mind-reading as a module. And here again, if it is, this work is quite a bit easier. The module is either there or not, it performs a specific function we can identify in behavior, and it emerged at some (relatively) determinate point in evolutionary history. But even if this is the right view of mind-reading, the boundary problem for cladistic grain applies: it would be hard to tell whether the homologous trait is a mind-reading system *itself* or some earlier-evolving system for social cognition that ultimately produced both human mind-reading and chimpanzee social cognition (whether mind-reading or not). We need some standard, some application of the criteria of homology, and perhaps the specificity constraint to compare against in order to tell the finer-grained homology from the coarser-gained.

But there are also good reasons to doubt the modular story of mind-reading. Plausibly, there are instead a number of different processes and neural systems that operate in social cognition in various stages and tasks. The ability to attribute mental states may not be an all-or-nothing capacity but a particularly sophisticated form of behavioral disposition attribution (see Chapter 1). If this is right, things get a lot murkier for the boundary problem. If many different systems are involved in social cognition, then the homology might hold between systems that are distinct from mind-reading (not merely differences of grain). If mind-reading is indeed a sophisticated kind of behavior disposition attribution, the mechanism responsible for chimpanzee behavior might indeed be the homologue to human mind-reading, but due to slight changes may not quite be capable of representing mental states.

So, in this example, the boundary problem is especially troublesome because the actual substance of the homology claim, the evidence for it, and the standards for evaluating the evidence are all unspecified. As such, it is very hard to know what to conclude. I take it that Sober leaves this unspecified deliberately, because he is more interested in what *follows* from this fact, and presumably wants the argument form to apply broadly. And again, he is clear that he only takes the argument to provide 'some' evidence for chimpanzee mind-reading. So, perhaps even building in this additional uncertainty, there is some small consideration of evidence here. However, if we start with weak evidence and weaken it substantially more, at some point it is no longer worth actually building into our evaluations of competing models (whether or not we should go that far is an open question).

The point is that this is how the boundary problem arises when homology attributions are underspecified. Unfortunately, when homology claims are *too* specific, we run into a different set of problems, apparent in the evolutionary psychology literature.

4.5 Evolutionary Psychology and Its Limits

The application of evolutionary reasoning to psychology finds its most ambitious forms in a subfield known as evolutionary psychology. As I frame it, evolutionary psychology aims to understand how psychological processes function *now* based on claims about what kinds of function would have been most adaptive in the evolutionary past. Evolutionary psychology, as a field of research, has mostly focused on human psychology, but the project can apply to nonhuman species as well (there have been calls for just that, e.g., Vonk and Shackelford 2013). The basic structure of the problem discussed here remains for any species.[18]

Evolutionary psychology (of this form) usually starts from a position of *massive modularity* (Cosmides and Tooby 1994, 2005). The general idea here is that the mind/brain is divided up into a number of distinct information-processing modules. Each of these has a specific task for which it is structured, and each runs independently of the others. This general view of cognitive architecture is contrasted with any view that posits something like general-purpose reasoning or learning (even if that view includes the possibility of domain-specific representations or knowledge). We can think of the massively modular mind as made up of independent, single-purpose computers.

The idea, then, is that each module solves a specific problem that our ancestors faced in their quest for survival and reproduction. So, the strategy is to identify a specific problem that our ancestors plausibly faced, posit a module that performs a task that would solve the problem, and then search for empirical evidence of that module. Perhaps the most famous example is the "cheater detection module." Cosmides and Tooby (1992) argue that one significant problem that would have needed to be solved is the enforcement of fair cooperation. If one individual fails to reciprocate on exchanges, that individual may gain the benefits of cooperation without paying the costs. Unless, that is, individuals in a group simply refuse to cooperate with an individual known to 'cheat.' Thus, they argue, it makes sense that a 'cheater detection module' would have evolved to track instances of social rule-breaking. They argue that the existence of the module is empirically supported by the fact that people better

---

[18] One possible difference does stand out: the lives that humans live now are likely very different than they were in our evolutionary past. We might think that the difference is less stark for most nonhuman species (as they live in the wild anyway), meaning that we might have a pretty good idea what their evolutionary past was like just from looking at the present. However, this would be a difference of degree, not kind, and should not be assumed in every case.

solve some reasoning tasks if they are framed as detecting social rule violation than as (formally identical) logical reasoning tasks.

As characterized so far, the evolutionary psychology project rests on two core claims. First is massive modularity. And second is adaptationism—the claim that current functional traits in a given species can be wholly (or substantially) explained by appeal to adaptation. In addition, Smith (2020) identifies a third central claim of evolutionary psychology: that it is able to identify what she calls *strong vertical homologies*. 'Vertical' homologies here is meant to specify that these are homologies reaching back into the past to identify traits in ancestral species, while homology is typically identified as a relation between traits in existing species ('horizontal' homology, as physically organized on the phylogenetic tree; Figure 4.3). 'Strong' is here meant to imply that the project requires that both the structures and the functions of each module are retained; so, it's a two-level homology that is stable across time at both levels. This is a difficult thing to show.[19]

Criticisms of evolutionary psychology have typically targeted its adaptationism, its massive modularity, or sometimes the combination of the two from various competing perspectives.[20] Smith's criticisms add the problem that the core of the view depends on homology assumptions that are vulnerable to boundary problems. The reliance on homology is hidden by the more obvious reliance on adaptationism. The traits present in a species at a given time constrain the direction of evolution, as does the environment by fixing what is adaptive, and these are often seen as competing explanatory posits. Since homology emphasizes the role of traits present in a common ancestor in how evolution unfolds, and adaptationism emphasizes the role of the environment, these might seem exclusive strategies. But evolutionary psychology actually requires strong versions of both, opening another source of uncertainty in its operation.

The version of the boundary problem that we encounter here is somewhat different from those discussed before, because the homologies are 'vertical' and 'strong.' The fact that the homologies are vertical means that we need to make claims about cognitive traits of extinct ancestors. The boundary problem, as we have seen, is hard enough when we have both species right in front of us, and having to work through the archaeological/fossil record adds

---

[19] Smith concludes that evolutionary psychology is 'impossible.' I don't go that far, though I do think this is a serious challenge for any research along these lines.

[20] As well as its troubling reductionism and genetic determinism; for a sampling of criticisms of evolutionary psychology see Samuels 1998; de Waal 2001; Panksepp and Panksepp 2000; Rose and Rose 2010; Gannon 2002; Buller 2006; Anderson, Richardson, and Chemero 2012; see also Gould and Lewontin 1979.

a substantial layer of difficulty. The fact that homologies are strong means that both the structural and functional features of the claimed module need to map onto one another. This, in effect, two boundary problems at once (admittedly, as noted before, the boundary problem for the functional homology may be harder than the structural one). It also requires that neither the structure nor function has shifted.

As in the last two examples, the modularity claim is doing a lot of work for evolutionary psychologists; indeed, it is doing a lot more work here. If the systems that evolutionary psychologists identify are indeed modular, it goes some way towards settling things. So, as before, it depends on how plausible we take the modularity claim to be. The modules that evolutionary psychologists posit are often very specific in the task they are purported to solve and, as such, in their function; note that the cheater detection module is just one module that may play a role in social reasoning and coordination. Even if we think that social coordination presents a set of problems that would need to be solved by any social animal, positing something as specific as a cheater detection module requires a certain conception of that problem space and a specific solution to the relevant part of it. Neither of these is guaranteed, and many other solutions are possible. More specific claims are, as a matter of pure probability, less likely to be true because they carry more commitments that all need to be accurate. In this context, this means that the view allows very little variation in structure or function across evolutionary time or individual learning and development.

Very specific claims about the modules mean that we can say a lot about the mind *if* the homology holds. But it also means that it is less likely that the claimed traits are shared. This is why evolutionary psychology runs into the opposite worry that faced Sober's cladistic parsimony argument. Sober's less specific, less detailed homology claims might be easier to support, but this is only because they carry less content. They are, thus, less informative. So, when we attempt to address the boundary problem by describing the homologies we take to hold, we face a balancing act which I call the *specificity trade-off*: go too far in either direction, and you pay a substantial cost.

Put more precisely, claims that some features of a trait in a comparison species can be expected in a target species require two inferences. The first is the inference that the homology holds (Inference 1 in Figure 4.4), and the second is the inference from the fact that homology holds to claims about the psychology of the target species (Inference 2). Any claim about a psychological capacity in one species based on (presumed) homology with another species is only as strong as this two-step inferential *chain*. The trouble is, as we have seen,

Fig. 4.4 Homology-based comparative claims require two steps: first, the inference that a homology holds, and second an inference about the target system based on that homology. There is a specificity trade-off in the respective strengths of these two inferences.

that attempting to strengthen one of these inferences by making homology claims more or less specific seems to necessarily weaken the other one.

Evolutionary psychology tends to make strong claims about how current traits function, thus aiming for a strong Inference 2. But this requires too much specificity in homology attributions and so comes at significant cost to Inference 1. On the other side of things, Sober's cladistic parsimony argument underspecifies the homology claim. This might allow a strong Inference 1 that *some* homology exists. However, it comes at the cost of a weak Inference 2—not being able to tell us much because we don't know the actual grain of homology that holds, or what is supposed to be included in the mind-reading capacity that is posited.

At these extremes, arguments fail because the utility of one of the inferences effectively goes to zero. Ultimately, we should aim for homology claims at whatever specificity best matches reality (that is, which features of the trait are actually conserved, and which not) as well as the task at hand. However, if we actually *knew* this much about the psychology of each species under comparison, we wouldn't need to make arguments about the relative probability of different hypotheses based on homology. In the meantime, it seems like the best goal is to aim for a moderate level of specificity that can support useful versions of both inferences, and conclusions that are reasonably plausible and reasonably informative.[21] So, for now, this is something we need to work with even as we try to get past it. I move on now to consider possible ways to get past it.

---

[21] As one way of imagining it: suppose that the utility of Inference 1 is 'u,' the utility of Inference 2 is '1-u,' and the utility of the overall chain is (Inference 1)*(Inference 2) or (u)*(1-u). This function has maximal value at u=(1-u)=0.5, and falls off to 0 as 'u' approaches 1 or 0.

4.6 Bootstrapping Homology: Integrating for Integration's Sake

Through the last three sections, I have discussed some prominent attempts to integrate neuroscience and evolutionary psychology into comparative psychology. These attempts, I argued, presuppose homology claims that they have not paid sufficient attention to. Too often, the relevant homologies are assumed, or their grain is treated as automatically understood. As a result, they run into trouble. The most important takeaway from each of these examples is that recognizing the challenges of homology should increase uncertainty in their conclusions. Stronger conclusions than this may be justified, but I am more interested in demonstrating the shape of the issue than fighting over any of these specific arguments.

Moreover, I take the basic worry here to generalize. Many attempts to integrate different fields into our understanding of animal minds will face these same challenges. For various reasons, and in various ways, they will require comparisons across species that require homology assumptions. Each of the three examples discussed above encountered the boundary problem in a different way, and so I expect others will as well. However, the lesson here should not be pessimism about integration. Rather, the lesson should be that integration needs to be attentive to homology claims as it requires them. Indeed, one important step will likely be to pursue *more* integration; integration of the right kind can help find and track homologies that can facilitate more global integration.

I have primarily discussed homology of cognitive capacities here, only addressing homologies in neural structure and behavior as they have become important. However, a more complicated, multilevel understanding of homology might be helpful in getting past the boundary problem and the specificity trade-off.

As a general matter, homologies might be identified in behaviors (as distinct from the capacities that generate them), or at various levels of neural structure, or in genetic or developmental mechanisms (which authors sometimes call *deep homologies*;[22] Shubin, Tabin, and Carroll 2009; Scotland 2010). Complicating matters, homology at each of these levels can come apart; the same structure might come to be produced by different gene networks, and the same behavior might come to be produced by different structural mechanisms (see Streidter and Northcutt 1991; Sommer 2008). However, as a first pass, we

---

[22] 'Deep' because the gene networks involved are often conserved across a much wider range of species, going back farther in evolutionary history than most readily apparent traits.

can assume they correlate enough that homology at one level is evidence of homology at another. Taking this approach, looking at different levels can provide evidence of homology.

For instance, suppose we are wondering whether there is a homology between the mechanisms responsible for similar-looking behaviors in two systems. If we discover similar features in the neural mechanisms, this might be evidence of the kind of *special quality* that is used to identify homologies. We might also find special quality in odd particulars of behavior. As noted earlier, biases that would be counterproductive to a task might be especially good evidence. This would mean that evidence of *failures* of nonhuman animals in tasks, if they look like human failures, could be the best evidence that their cognitive capacities are like ours.[23] It might also sometimes be that we can identify anatomical features such as brain structures or network architectures that we can trace across a larger number of species (assuming that the anatomical features are easier to observe than cognitive capacities). This might provide evidence of *continuity of intermediates* as well, further supporting a homology claim.

Along these lines, de Waal and Ferrari (2010) argue for a 'bottom-up' approach, over the common 'top-down' approach that studies behavior first. They suggest that studying the 'nuts and bolts' of cognition, the basic neural mechanisms, might provide evidence of homology between humans and nonhuman animals, and thus evidence of similarity. One of their examples is mirror neurons. Mirror neurons fire both when performing a specific action and when observing others performing it. De Waal and Ferrari cite evidence of activity by mirror neurons in humans and macaques as evidence that 'imitation' in apes and monkeys is much like imitation in humans. However, the mere fact that humans and monkeys both have mirror neurons does not tell us much about actual imitation abilities, or what role they play in their lives (Spaulding 2013; and, as de Waal and Ferrari note but dismiss, the evidence about which monkey species actually do imitate is controversial). A thinner interpretation treats them as evidence of special quality that might be shared between the human and monkey mechanisms for imitation, and thus as evidence of homology. This does not itself answer questions about imitation, and the nature and grain of the homology are yet to be established. However, it might encourage new ways of gathering and organizing evidence.

---

[23] In effect, this could be a non-circular way of filling in Sober's claim of behaviors being 'the same' from his cladistic parsimony argument.

120 INTEGRATION AND HOMOLOGY

Behavioral genetics and developmental psychology can, in principle, provide another angle on the question. That is, if we see similar gene networks underlying some psychological capacity across species, and/or similar developmental trajectories across species, that might also be evidence of homology. However, these fields have a long way to go before they can contribute much directly. Behavioral genetics has to work not only through the thickets of complexity described in this book but also through complexity in genetics itself (e.g., Nelson 2018). As such, it typically studies global traits like personality, or psychiatric disorders such as alcoholism or schizophrenia, rather than specific cognitive capacities. Similarly, developmental psychology has focused almost exclusively on human development and studied development in nonhuman animals only very little (for calls that each of these fields move in each of these directions, see de Geus et al. 2001 and Rosati et al. 2014, respectively).

The boundary problem still holds at each of these levels also, of course. And if we are working at multiple levels, we face the additional issue of getting those grains to line up. However, having multiple targets at different levels might give more sources of evidence. Overall, the hope would be for a virtuous kind of bootstrapping, where multiple threads can be tightened slowly, each providing some support for homology claims. These homology claims would scaffold better cross-species comparison, in turn producing better understanding of homology. This can repeat in a virtuous cycle.

4.7 Conclusion

Today, as in 1978 when the Sloan Report was written, the future of the science of animal minds is one of increasing integration between the various subfields that look at different parts of animal cognition in different ways to different ends. This integration is important; no single source of evidence or type of experiment or observation will be able to settle the question we most want answered about animal minds. This has been the theme of this book, if anything has. We must look to sources of evidence wherever we can find them.

Integrating evidence, taking evidence from one science and applying it to evaluations of claims in another, is difficult work. In particular, I have argued here that attributions of homology play a central role in bridging these gaps for many such attempts. The significance of homology in this role has been underappreciated. Indeed, it has too often been ignored or assumed, to the detriment of attempted integrations. Attempts to integrate neuroscience into cognition will require homology attributions when the data or theory from neuroscience was developed using a different model species from the species currently under consideration. Attempts to integrate evolutionary theory into cognition will require homology attributions whenever there is a comparison of species based on relatedness, whether those species are current or one is ancestral. These are not the only ways that integration can be attempted, but they account for a large portion.

Researchers looking to help gather evidence for psychology must be alive to the role that homology may play in their inferences and the challenges that can result—in particular, the *boundary problem* and the *specificity trade-off*. The answer, then, is not to resist or slow integration; it is for integration to be attentive to homology, and even to aim towards identifying homologies that can scaffold further integration. When it comes to any specific claims, the challenges discussed here largely mean that conclusions based on uninterrogated homology assumptions should be weakened and hedged even further. But this is not an insurmountable problem for integration, and it is not reason not to pursue it. Indeed, it is reason to pursue integration more thoroughly and systematically than ever.

5

Ecological Validity

5.1 Introduction

Imagine you are picked up by aliens and flown into space for tests. They present you with objects you have never seen and apparent tasks with purposes you don't understand. As you try to figure out what to do, you have no way to communicate with your captors; they bustle around speaking their alien language. The chamber you are in seems to defy your sense of space, and you are overwhelmed by blinding lights and unfamiliar smells. You are hungry and confused, so you poke around hoping you can figure out how to get home, or at least get some food. This experience, extraterrestrials aside, may not be all that different from the experience of many animals the first time they are presented with an experimental laboratory task. Laboratories, compared to the wild, present unnaturally shaped rooms that are unnatural colors with unnatural lighting and unnatural smells and sounds. Experiments require interaction with (or are in the presence of) humans and artificial materials they would never encounter in the wild, to perform tasks that may be irrelevant to their natural way of life.

Researchers who worry about this artificiality typically phrase those worries in terms of *ecological validity*. At a first pass, ecological validity is the degree to which artificial contexts accurately recreate contexts in the wild. Loss of ecological validity, to at least some degree, is often seen as the cost of doing business in controlled laboratory settings. Artificial situations and materials allow greater experimental control over the innumerable variables that might impact cognition. This greater control allows greater confidence about which parts of the environment the animal is responding to and, thus, the causal structure of the cognitive processes involved. As such, there is generally a trade-off built into the laboratory study of cognition: with increased control usually comes reduced ecological validity. This is arguably a feature of science in general: many biology experiments are done in petri dishes, and physics experiments done in vacuum chambers.

Seven Challenges for the Science of Animal Minds. Mike Dacey, Oxford University Press. © Mike Dacey 2025.
DOI: 10.1093/9780198928102.003.0006

INTRODUCTION   123

Different approaches to this trade-off also mark disciplinary differences within the study of animal minds. This has most obviously marked the difference between the fields of *comparative psychology* and *cognitive ethology* (as well as ethology and ecology more generally; Allen and Bekoff 1999). Comparative psychology generally studies animal minds using experiments, usually in laboratories. In contrast, cognitive ethology generally studies animal lives in the wild. While they overlap, the two often have different research foci, methods, and interests. One of the goals of this chapter will be to scaffold the integration between these two fields.[1]

Unfortunately, explicit discussion of ecological validity in the literature (across philosophy, various subdisciplines of psychology, and ethology) has mostly occurred as part of discussions of single experiments. That is, discussion sections of empirical papers often admit, as a possible limitation, that the reported experiment fails to be ecologically valid. Or, alternatively, a critic of a specific experiment may point out its lack of ecological validity. Perhaps the main reason that so little has been said about it as a general topic is that it is hard to say anything about what, in general, makes studies ecologically *in*valid in the important ways, and as such how to correct or interpret them. There are myriad features of any given task or context that might have unexpected or unpredictable influence. Nelson (2018), in her anthropological case study of behavioral genetics research at the pseudonymous "Coast University," describes levels of care and control among behavioral experimentalists over any possible confound that might appear fully superstitious to outsiders. Researchers are unwilling, for example, to change deodorants or grooming patterns in the middle of a weeks-long period of data-gathering for fear that any change in smell may 'break' the observed behavior in the study mice. This is, as it turns out, reasonable enough; the number of features of task and context that may matter is innumerable, and we simply don't know which do matter until we try.

I won't be able to say anything concrete about those features or how to identify them. Instead, my aim is to put some general shape on the problem of unnaturalness of laboratory studies and its practical implications. The biggest step is to downplay the question of whether *individual* experiments are ecologically valid (as is the theme of the book); looking instead at relationships *between* experiments allows us to build bridges between wild behaviors and laboratory tests.

---

[1] I have mostly focused on comparative psychology throughout the book, because of my background in thinking about experimentation and cognition. This should not be taken as a lack of appreciation of the value of ethology and ecology. In fact, I think that a comparative psychology that takes the advice of this book would look quite a bit more like cognitive ethology than it does currently.

5.2 Two Experiments with Bees

For a concrete sense of what it means to have a more artificial or more natural experiment, I present two experiments on bee cognition. The first demonstrates a particularly artificial experiment, while the second is more natural. Each of these has its strengths and weaknesses.

In the first example, Howard et al. (2019) argue that honeybees are capable of addition and subtraction. In this experiment, bees were trained on a Y-shaped maze, one arm of which contained a sugar-water reward, and one arm contained an aversive quinine solution as a punishment. As the bees first entered the maze, they saw a sample stimulus which included a particular number of elements (one, two, four, or five shapes that were squares, diamonds, triangles, or circles) in either blue or yellow (see Figure 5.1). Once in the maze, past the sample stimulus, each arm was labeled with another set of similar elements of the same color in the crucial stimulus. The task was this: yellow was used to indicate that rewards would be found in the arm labeled with one *fewer* element than the sample display, while blue indicated the arm labeled with one *greater*. In other words, yellow required subtraction of one element, and blue required addition of one element. The stimulus on one of the arms would be a correct answer (either one more or less than the sample), and the other arm an incorrect answer.

Howard et al. trained the bees on three numbers each for addition and subtraction, and in the key test the novel number 'three' was used for both; thus a correct answer for addition was four, and for subtraction was two. In this transfer test, they found that bees chose the correct path significantly more often than chance in all four variants: for addition and subtraction each, they included examples in which the crucial number of elements differed in the same direction and the opposite direction of the right answer (e.g., for subtraction the incorrect crucial stimulus had more elements than the sample in some trials, and fewer than the sample in others).

Their claim that this demonstrates addition and subtraction in bees is controversial. Shaki and Fischer (2020) note that the correct answers in the transfer test were also the closest to the average number that had been reinforced in training; for example, the correct answer of 4, in the addition variants, was closer to the average number reinforced in addition tests (3.33) than were the alternative options (5 and 2). They suggest that the bees may have simply tracked this similarity. MaBouDi et al. (2021) also ran experiments with several other controls, and suggest that bees could accomplish the task by tracking

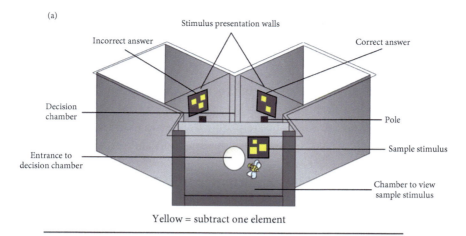

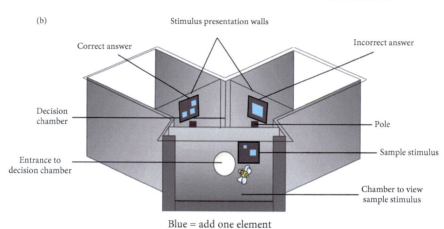

Fig. 5.1 The experimental setup from Howard et al. (2019). Reprinted from Scarlett R. Howard, Aurore Avarguès-Weber, Jair E. Garcia, Andrew D. Greentree, and Adrian G. Dyer, "Numerical cognition in honeybees enables addition and subtraction," *ScienceAdvances* 5(2), © 2019 The Authors, distributed under a Creative Commons Attribution NonCommercial License 4.0 (CC BY-NC).

non-numerical properties of the stimuli like spatial frequency or grain in the stimuli.

The point for now, though, is that most of this experiment is artificial. The Y maze is an unnatural setting, the numerical stimuli are unnatural, the sugar-water and quinine reinforcers are unnatural. The task itself is likely unnatural; it's not clear what role addition and subtraction would play in a

126   ECOLOGICAL VALIDITY

bee's life, or at least, what role it would play in feeding.[2] Arguably, the experiment is hard to interpret because of these artificial features; it's not clear how the bees themselves interpret the task, and it's not clear what they see in the artificial stimuli.

As a more naturalistic example, an experiment by Gould and Gould (1982), later replicated by Tautz et al. (2004), tested the cognitive abilities involved in their 'waggle dance.' A worker bee returning from a successful foraging trip will 'dance' in a pattern that demonstrates the direction and distance to the food source. Potential recruits will observe the dance, including the speed of the movements which correlates with the quality of the source. They also smell the pollen on the dancing bee, and then decide whether to follow those directions. If they do, they can fly directly to the source, perhaps a patch of flowers that recently bloomed, even if the source is over 2 miles away. The waggle dance has been seen as the prime example of sophisticated communication in insects since it was first discovered (von Frisch 1974).

Gould and Gould (1982) and Tautz et al. (2004) aimed to test how recruit bees understand what they learn from the dance. They trained bees to visit a feeder that was placed on a boat in a lake, as well as a feeder placed on land, equidistant in the opposite direction. Later on, the researchers found no recruits that had followed the dances of trained bees to the lake feeder, but many to the land feeder. They hypothesized that the potential recruit bees had rejected the dance as having indicated an 'implausible' location over open water. If true, this would indicate a substantial level of understanding on the part of recruit bees, perhaps even a mental map against which an imagined flight plan could be compared. However, using newer tools, Wray et al. (2008) ran a version that observed behavior of the bees inside the hive. They compared video of potential recruits watching waggle dances from a lake feeder and a land feeder, and found no difference in the number of recruits who left the hive after watching dances for each location, and no difference in the time they took to leave (e.g., no sign of hesitancy to follow the lake dance). They also found no difference in the number of recruits that arrived at each feeder. They speculate that the results of Gould and Gould (1982) and Tautz et al. (2004) might have been due to bees failing to find the lake feeder. The visual effects of flying over

---

[2] If there is such a capacity, it may be more useful in hive 'politics' such as the collective decision about where to go during a swarm. After the swarm has left the old hive, scouts who have found potentially suitable locations will perform a waggle dance. If there are competing candidates, the decision is made based on which candidate is able to slowly accumulate the most dancers (Seeley 2010).

water can interfere with bees' distance estimations, perhaps disorienting them. Additionally, Tautz et al. used unscented feeders. Bees generally use scent to find the exact location of flowers once they have gotten close, so this robbed them of another navigational cue.

While the more exciting result seems not to be borne out, I am here, as in the last example, mostly interested in the strategy. The bees lived in a hive, and the artificial feeders gave out sugar rather than pollen, but the overall setting would have been pretty similar to life for bees living in a hive in rural farmland. The task involved natural feeding behaviors, and the main manipulated variable, location of the food source, is one the bees would encounter naturally. Despite the advantages of this naturalistic setting, it also has downsides: the problems with the original lake experiment arise because of a lack of control. It was clear that bees were not showing up at the lake feeder, but it was not at all clear what caused that.

Overall, these two experiments illustrate some of the ways an experiment can be more or less artificial, and some of the risks that come with each. More generally, bees make a good case study for ecological validity because, even when 'domesticated,' they are still in many ways wild. Bees living in a standard hive roam and forage as they please, but still return to the hive. Moreover, in experimental settings, the bees pay attention to what they will; it is harder to direct attention than with larger mammals (though not impossible). There is also more practical interest in bee foraging because of its importance for agriculture. For these reasons, there is a substantial base of research on the topic that will ultimately play an illustrative role in my proposal (Section 5.7). First, though, we will attempt some general lessons from existing literature on ecological validity.

5.3 Ecological Validity: The Basic Issues

5.3.1 Initial Orientation

While ecological validity has received disappointingly little critical discussion as a general topic in the comparative psychology and philosophy literatures, it has received some in developmental psychology and education research. Even in those discussions, though, there is no consensus on a definition of ecological validity, or on the key relationship between experiment and environment. Though the project of producing a formal definition of ecological validity is both valuable and underserved, I take a more practical orientation here. I aim

128   ECOLOGICAL VALIDITY

to identify and address the key *practical* problems that might arise due to lack of ecological validity; what mistakes might we make, or what inferences might we be prevented from making, because of the 'unnaturalness' or 'artificiality' of laboratory experiments.

Taking this orientation, problems of ecological validity are united by a shared cause of 'artificiality' rather than any formal features. Sometimes, these problems will manifest as a subspecies of 'external validity.' External validity is about generalizability of results; the applicability of a study to anything outside that particular laboratory context (e.g., the wild). Sometimes problems will manifest as a subspecies of 'internal validity.' Internal validity is about whether the experiment actually measures what it is intended to measure. This occurs when unnatural features of the experiment lead the animal to interpret the task differently than the experimenter intends. When this occurs, the animals might not be using the capacity that the researcher intends to study.

There is an asymmetry in the way that considerations of ecological validity are usually taken to apply to different experimental results: negative as opposed to positive. Experiments in comparative psychology are often set up such that they test whether an animal is able to 'succeed' at some task. When animals do not succeed at the experimental task, it is a negative result. Negative results, in general, are very difficult to interpret because there are so many ways an experiment might go wrong. More specifically, there are a number of reasons why an artificial context may cause an animal not to perform up to a level at which it is capable. Animals may not be motivated to engage with an artificial task. They might be confused by the task or context, or they might simply interpret the task differently than intended by the experimenter. For these reasons, the artificiality of experiments makes it hard to conclude anything from negative results, as they are so easily explained away by appeal to artificial features of the experiment.[3]

Different worries apply when the experiment seems to display animal success. An animal may 'succeed' by the behavioral standards of the experiment, but without recognizing it as the kind of task it was intended to be—that is, they might solve it using some other capacity than the intended target (trial-and-error learning or associative learning might be the most common possibilities). For instance, MaBouDi et al. (2021) argue that the bee arithmetic

---

[3] For comparison, you might reasonably object if you somehow find out that the aliens in our opening thought experiment concluded that you need not be granted any moral or agential status based on your 'failures' in their tests.

experiment may have gotten the results it did because the bees were tracking something other than addition and subtraction: they do not "see" the task in the same way the experimenters intend. In general, experimental setups can undergo long periods of development aimed at getting them to "work" by the standards of the study, not to make them natural (e.g., Sullivan 2022). It's also arguably possible that behavior is caused by the target capacity (e.g., mind-reading or causal cognition), but the artificial setting causes that capacity to operate in such an unusual way that we cannot generalize the theoretical conclusion. Even if the bees did perform arithmetic in the Howard et al. (2019) experiment, this may not tell us anything about how they behave in the wild.

We might think, then, that anyone aiming to dismiss an experiment demonstrating animal success on grounds of ecological invalidity might owe an explanation of how the animal did succeed in the behavioral task, but anyone aiming to dismiss an experiment showing animal failure might not owe any such explanation. I think it's right that there is an asymmetry along these lines, but we should be careful about hastily rejecting *all* negative results. Negative results can tell us something, especially if there is persistent and systematic failure in multiple experiments on related tasks.

Authors raise ecological validity as a worry across branches of psychology. But like most of the seven challenges, it is perhaps especially troublesome in comparative psychology. Adult human participants (usually, Western college students) do live much of their lives in rooms like a lab, and do interact with the kinds of artificial materials, including computers, that an experiment might include. An adult human can also be instructed about what they should be doing and why (if it serves the goals of the experiment). It's also more likely that adult human experimenters are able to design materials and tasks that elicit the intended cognitive processes and behaviors in other adult humans. None of this is true of work with animals. Work with infants and toddlers in developmental psychology arguably falls somewhere between these. In any case, I take the core practical issues to be the same, even if they differ in degree. Thus, my discussion will draw from work addressing ecological validity in various branches of psychology.

In several senses, then, I am taking a capacious orientation to ecological validity. I consider ecological validity to be a reasonable subject of concern in all settings in which artificiality might trouble interpretation. This use of the term may crosscut terminology employed by other authors. However, there is something of a terminological quagmire here that I would like to avoid (e.g., Schmuckler 2001; Araújo, Davids, and Passos 2007; Kihlstrom 2021).

130 ECOLOGICAL VALIDITY

I adopt this capacious understanding because I think it best serves my pragmatic orientation to the issues. With such an approach, one can go a long way toward addressing the practical problems without settling the formal or definitional issues. The fact that these problems are so deep and multifaceted is perhaps partly due to the disappointing lack of explicit attention that has been paid to understanding ecological validity. I also take my capacious use of the term to fit the practical ways it is applied when it comes up in discussion of individual experiments. When ecological validity is raised as a limitation or criticism of a specific finding, basically any unnatural feature of an experiment is fair game. I see little reason, at this stage, to carve up these worries for formal or definitional reasons.

We need not always be worried about ecological validity. Some experimental aims, discussed more below, simply don't need naturalness (and may encourage unnaturalness). Proof-of-concept and exploratory studies may be unconcerned. In other cases, a task may be well validated in previous work such that we need not take these concerns as significant. But where ecological validity is a concern, it would be helpful to have a more systematic way of thinking about it than we do currently.

Stepping back a bit, we can think of worries about ecological validity, where they apply, as widening the inferential gap between an experimental result and theoretical interpretation of those results (discussed in Chapter 1). There are many factors in experimental design and setting that might be unnatural in ways that unexpectedly influence behavior. This is one way of saying that we might not know what is actually causing behavior, and, thus, is one reason it is hard to interpret experimental results. The goal in this chapter is to find ways to minimize the contribution that naturalness makes to this inferential gap. If we can shrink it, even if we cannot close it entirely, that is progress.

5.3.2 Dimensions

Even if we don't have a single general account of what kind of artificiality matters, we can say some things about the kinds of artificiality we are likely to encounter. Schmuckler (2001) provides a framework by organizing existing treatments into three 'dimensions' of ecological validity (in developmental psychology). I summarize these here. This barrage of artificialities might make things seem quite dire, but I don't think that it must. Which of these actually matter will vary case by case.

5.3.2.1 Dimension 1: Nature of the Research Setting or Context

The lab itself is an artificial place. It has artificial walls and barriers; perhaps organized in a Y maze. Experiments with smaller animals are often performed in cages. Lighting is artificial, often set for human comfort, as is climate control. The smells of a laboratory are very different from those of the wild, which can matter significantly for smell-dependent animals. Sometimes animals are allowed to join the experiment voluntarily (e.g., usually, apes), but other times they may be artificially forced or cajoled into performing a task; perhaps they are deprived of food as a motivator to participate in an experiment that uses a food reward. These imposed motivations override other motivations or behavioral patterns that might surround natural foraging and feeding. Animals may be physically restrained. They are often in the presence of humans, if not interacting with them directly (as noted in the worries about personal hygiene and smells above; Nelson 2018). Animals naturally live in the woods or on grasslands, in burrows or on reefs. Each of these has its own ambient lighting, smells, and sounds. And these elements of the seeming background may contain subtle cues that we miss, which are significant to the animals. They also initiate behaviors freely as part of natural patterns and rhythms of life. Humans are often (hopefully, anyway) not physically present, looking over their shoulder in their natural lives. It's not always clear how unnatural settings can impact behavior, but it is always possible that they do.

5.3.2.2 Dimension 2: Nature of the Stimuli

The stimuli and materials used in an experiment are often artificial, and often intentionally chosen to not be naturally significant to the animal. Simple cues like a light bulb or a tone are common in paradigms like classical conditioning, pairing a light with food—a holdover from behaviorist research.[4] Other experiments require interaction with plastic toys or artificial tools of various kinds. Some even use computer screens, either simply watching or interacting with touch screens. In the bee experiments described above and below, artificial flowers are sometimes used, as are artificial scents and sugar-water (rather than nectar) reinforcers, and artificial patterns intended to indicate number in the arithmetic studies. These may have a different significance to the animal than the researcher intends (and the experiment requires).

---

[4] This is one area where the stimulus is often intentionally artificial, in hopes that it doesn't have any significance for the animal before the experiment.

132 ECOLOGICAL VALIDITY

5.3.2.3 Dimension 3: Nature of the Task, Behavior, or Response

The behavioral response that an experiment measures can be itself artificial, or it can have an arbitrary relation to the stimulus materials. Classic responses in comparative psychology include things like pressing a lever (though see Timberlake 2002; Section 5.7) or indicating a location to a human experimenter. Materials using computer screens might require interacting with that screen. As a general matter, food rewards are often used (on the assumption that food is a more-or-less universal reward), but the behaviors usually have little to do with natural feeding or foraging. This is the case in the bee arithmetic experiment as well; it's not clear what role arithmetic would play in their natural foraging behaviors.

5.3.3 Animal Backgrounds: Upbringing, Socialization, and Training

Schmuckler's dimensions focus on the conditions of an experiment itself. However, the background of the animals being used is a major source of artificiality as well. The animals used in laboratory experiments are usually raised in captivity, meaning they grew up in an artificial environment with artificial (at best) social arrangements and extensive interactions with humans. The influence of this history will vary greatly by species. For bees, it may not matter much. But for other animals, the difference can be significant. Rats and mice might be housed in small containers, as a small group or alone. Laboratory rats and mice also belong to genetic strains that have been long bred apart from their wild cousins. The genetic makeups of these strains are well known, but they would not survive in the wild. Apes may have a reasonably naturalistic living area, but they will be extensively socialized to humans, and may be more directly dependent on their human caretakers than other apes they live with. As the captive care of apes for experimental purposes has been banned and/or defunded, the apes in experiments now are often in care because they wouldn't survive in the wild either, which may be interpreted as more reason to doubt their behaviors are representative of their wild relatives.

Along the same lines, I would also like to raise a feature of many laboratory experiments that has been underappreciated in this literature: the large number of training trials that are often required before the experiment can even be run (sometimes hundreds or thousands). This may often be an effect of the dimensions of ecological validity described above: the artificial task means that the animals must learn from scratch. However, reinforcement training is itself an unnatural feature of the experiments which plays an important role in the animals' behavior. This means that the animals, once they encounter the actual experimental trials, are more different from their wild relatives.[5]

The unusual life histories and training of these animals add another layer of artificiality to experiments. This worry might seem different: the worry is whether the individuals are representative of their species, not whether the experimental task is natural. Training, though, blurs this distinction. It is, in a sense, part of the experiment, and yet it is also in the background of the parts of the experiment that are usually analyzed. All of these worries fit with my capacious orientation toward ecological validity, though. The background of the animals may be worth including as a fourth dimension of ecological validity. Moreover, I think the first steps toward a solution to all these problems are the same.

5.4 Lab and Field: Across the Great Divide (?)

The role of ecological validity in the study of animal minds is shaped by, and in turn shapes, the disciplinary boundary between comparative psychology and cognitive ethology. Comparative psychologists operate in a laboratory setting, designing experiments for animals to perform. Cognitive ethologists work in the field, observing natural behaviors and interactions. These, of course, require very different methods and research questions.

Laboratory psychology leverages the control of the laboratory setting in order to make more confident claims about cognition. Field research, on the other hand, gives a window into the actual lives of the animals. This typically involves less hyperfocus on cognitive mechanisms and more interest in context. 'Context' here includes environmental context, such as the set of challenges the animals face and the kinds of problems they need to solve in day-to-day life. It can give us social context, about how groups are organized, and how an individual's position in the group might influence behaviors and relationships with specific others. Ethologists also tend to be interested in evolutionary context, about which behaviors would have been adaptive under

---

[5] One option might be to argue that these experiments ought to be seen as tests of the animals' capacity to learn these tasks. However, usually no data relating to training is analyzed (only recently has it become common to report any data about training at all, usually in supplemental materials), and the major manipulations of these tests are not manipulations of learning context or strategy. There is something to the thought that animal experiments test the flexibility of cognition (see the next section), but such a reinterpretation of the experiments can't itself solve worries about artificiality.

which circumstances, and what purpose various behaviors might serve in order to have evolved. Regarding specific behaviors, field research can help tell us which behaviors are actually significant to the animal and affords the chance for unexpected observations. It allows the animals to demonstrate their thinking to us, rather than forcing them to perform to a human-designed experimental task.

The two fields have tended toward different theoretical orientations as well. In broad terms, cognitive ethologists have tended to be more willing to posit and study consciousness in nonhuman animals, and have not viewed anthropomorphism as a major problem, or have even viewed it as a valuable scientific tool. On the other side of the divide, comparative psychologists have been more likely to shun discussion of consciousness and treat anthropomorphism as a cardinal sin. This difference is sometimes marked by the distinction between 'romantics' and 'killjoys.' Where 'romantics' attribute humanlike psychological abilities of all sorts to nonhuman animals, 'killjoys' deny them, or at least express skepticism (Dennet 1983). While romantic versus killjoy arguments certainly play out within each discipline, on the whole comparative psychology is comparatively more aligned with killjoy approaches, while ethology is comparatively more aligned with romantic approaches. When the difference in theoretical approaches does align with the disciplinary divide, things can get contentious (see Bekoff and Allen 1997). Arguably, tensions have eased in recent years, and the fields have grown closer together (Andrews 2020), but the task of integrating their approaches and findings remains a difficult one.

It might be tempting to conclude that these fields have sharply divided aims. There is, admittedly, something to this. I'd even suggest there are at least three general questions that individual studies on animal minds might have as their primary focus:

Lives: What are their lives like?
Mechanisms: How is information processed in their cognitive mechanisms?
Limits: What are the limits of their cognitive/behavioral abilities?

Questions about their lives would be the straightforward purview of ethologists and ecologists, while questions about mechanisms would be the primary purview of comparative psychologists. It is worth saying a bit more about questions about limits of cognitive abilities.

As noted, cognitive ethologists have generally been more willing to attribute intelligence to animals and may study these capacities at their best, used as they were selected for in the wild. However, laboratory experiments also often aim to demonstrate the limits of performance on some task.[6] The Howard et al. (2019) bee arithmetic study has this orientation, attempting to test the limits of bee cognition regarding numerosity: is it sophisticated enough that they can do arithmetic? When this is the aim of a study, it changes how we think about artificiality. In fact, researchers sometimes take an unnatural task or behavior to *benefit* experiments probing limits because it is reason to think the observed behavior is truly novel. For example, Loukola et al. (2017) demonstrated that bumblebees were able to learn to roll a ball onto a target for a food reward, simply by observing other bees doing so. Observer bees would even improve upon the methods they observed, for example pulling rather than pushing the ball, apparently demonstrating flexibility and recognition of the action as goal-directed. The authors emphasize that this task is far removed from the bees' natural foraging behaviors, and so the experiment cannot be explained away as a simple modification of existing behaviors: the bees are learning something genuinely novel.[7] As such, the authors argue that the experiment "probe[s] the limits of behavioral and cognitive flexibility in bees" (833).

These differences in aims, along with the differences between field and laboratory study, might be taken to imply that these fields can't and shouldn't *want to* integrate. This response would be too hasty. First off, there are non-evidential kinds of integration which are uncontroversial. Most obviously, each field may raise specific questions that others may attempt to answer. For example, field research can identify new behaviors that we can, in turn, study in the lab. Finding ways to integrate *evidence* from these different research programs is more difficult, but still important. I have argued that individual experiments and observations will only provide weak evidence, so it is important to pull from as many sources as possible.

There are, thankfully, some clear ways in which answers to each of the three questions can inform answers to the others. Understanding animal lives in the

---

[6] Often without speculating on the mechanisms responsible. Here we come back to the discussion of modeling and Allen's (2014) complaint about 'trophy hunting' discussed in Chapter 3. Note also that the distinction between the study of limits and study of mechanisms gets more complicated when we address the role of learning (the mechanism for reaching limits), especially in light of the backgrounds of the animals (Section 5.3.3). The next chapter will discuss a 'limits' experiment in detail (Inoue and Matsuzawa 2007).

[7] I am wary of any overly strong distinction between 'truly novel' behavior and simple modification of species-typical behavior. Any such distinction risks the features of the associative/cognitive distinction that I criticized in Chapter 3. Most often, I suspect, behaviors fall somewhere between these two extremes. I'll note here that bumblebees dig and maintain nests, which requires moving things around. But these behaviors are not naturally connected to the food reward, so that may be novel. Overall, though, I do take this kind of approach to be valuable in at least some cases.

136   ECOLOGICAL VALIDITY

wild can provide context to claims about mechanisms. That is, it can help researchers understand the tasks that the processes actually need to solve, and the contexts in which they are used. Understanding the limits of a capacity can also determine which hypotheses about mechanisms are plausible: some candidate mechanisms cannot plausibly be expected to perform over a certain level. Additionally, understanding mechanisms can inform understanding of the animals' lives because it can inform interpretation of observed behaviors and the intentions behind them. These piecemeal integrations are valuable, but it is worth asking whether there is a more systematic framework to help with this integration, as well as helping address the other problems of artificiality noted above.

Summing up, problems with artificiality can arise because of artificiality of the experiment itself (along any of Schmuckler's [2001] three dimensions) and the backgrounds of the animals, including training. They can, in turn, cause problems with external validity, internal validity, and integration between research from the lab and the field. I will sketch a framework for thinking about ecological validity that I think can help address these worries. To set that framework up, I describe two influential discussions in the existing literature: first Mook's (1983) defense of external invalidity, then Brunswik's (1943) introduction of ecological validity to the psychological lexicon.

5.5 "In Defense of External Invalidity"

Even optimists about laboratory studies generally concede that concerns about ecological validity are often important. In contrast, Mook (1983) provides a rare, spirited defense of studies that *lack* external validity. Mook discusses external rather than ecological validity, and he is actually discussing work on human cognition (his area). Even so, his discussion transfers to concerns about generalizability of unnatural or unrepresentative experiments with animals, so I will treat this as a defense of ecological invalidity as well (up to a limit we'll encounter below). Specifically, Mook criticizes what he sees as a tendency to dismiss or discount experiments immediately upon noting unnatural features, decrying what he calls the "count 'em mechanics" of simply listing the differences between an experimental study and real life to discredit the study.

In fact, he argues, external *invalidity* can be an important feature of a study, depending on the goals. Sometimes the goal of an experiment will be to study effects that can be immediately applied in the world (he calls this 'the agricultural model'). This includes, most obviously, work on humans that applies directly to education or politics. Work in education should be directly generalizable because researchers want to develop teaching methods that will work in schools, and work in politics needs to be generalizable because we want polls to reflect the attitudes of the voting population. Researchers with such an 'agricultural' goal should be deeply concerned about criticisms of external validity.

However, Mook argues that many experiments have a different goal: to test a prediction that is made by a theory about a particular context. That is, a theory predicts a certain behavior in a certain task/context, and the experiment is designed to see if that prediction is accurate. If it is not, the theory should be modified or abandoned. In these types of experiment, the goal is not to generalize *results* at all. Instead, *theories* are generalized. What he describes here fits most obviously with what I termed work on cognitive mechanisms in the last section.

So, the idea goes, an experiment can bear on the theory about cognitive mechanisms by confirming it, disproving it, or requiring that it be modified. The resulting theoretical conclusion is what is generalized. This means that, firstly, the result itself doesn't have to be generalized in order to bear on the theory. And secondly, it is exactly in virtue of their (intended) generality that theories should apply even in artificial contexts. Mook notes, "the *processes* that we dissect in the laboratory also operate in the real world" (385).

Thus, he distinguishes between those studies that aim to determine what 'does' happen (in the real world) and those that are interested in determining what 'can' happen (in a specific set of circumstances set by the experimenter). Those experiments interested in determining what 'does' happen are subject to simple criticisms based on external validity, while things are more complicated for those interested in what 'can' happen. He concludes that researchers need to think through, case by case, "(a) what conclusions we want to draw, and (b) whether the specifics of our sample or setting will prevent us from drawing it" (386).

Mook is right to reject "count 'em mechanics" in favor of a more careful consideration of the factors involved. He is also right that different kinds of studies will have different ends, which require different kinds of generalizability. However, at least for our purposes, we can't go quite as far as he does. First, notice that generalizability is not the only worry that results from unnaturalness; unnatural settings can potentially mean that the animal is doing things in a different way than the researcher intends (that is, it can threaten *internal* validity as well as external validity). As such, even if the processes we "dissect in the lab" operate in the real world, we cannot always be sure which we are dissecting. Second, while his distinction between studies of what 'can'

138   ECOLOGICAL VALIDITY

and 'does' happen might appear to map onto the difference between 'mechanisms' and 'lives' from last section, the 'agricultural model' is much too narrow to capture most study of animal lives. His examples describe research that is interested in making concrete predictions about the results of a single intervention or measurement in the world (a vote, an educational method, etc.). However, an understanding of animals in the wild often cannot be reduced to such a specific question. Understanding life in the wild requires how animals might respond in a number of naturally occurring conditions across a population and a lifespan.

Despite these differences, the general structure of Mook's framing of 'can' experiments is instructive. Instead of taking the *results* to generalize, he inserts a step between the experimental result and generalization. This step is the evaluation of the theory under consideration. The insertion of some such step is an important insight, but I don't think the direct evaluation of a theory based on the single experiment is the right way to conceptualize it. I have argued (in Chapter 1) that most individual experiments in comparative psychology can't bear on the theory this directly. Another influential discussion from the literature will help us think about this specific issue.

5.6 Brunswik and Compiling Experiments across Situations

Brunswik (1943) coined the term 'ecological validity' and provides one of the most influential historical discussions of it.[8] For our purposes, his discussion will help us think about how to manage these issues when a single experiment is not doing all the interpretive work. Brunswik notes that, for any environmental feature that an individual (human or nonhuman) may need to understand, there are a number of specific cues they might use. As his example, he discusses visual recognition of depth in humans. There are a number of cues that the visual system can use to detect depth, including, for example, binocular parallax, occlusion by closer objects, and atmospheric effects. Each of these cues may contribute to the perception of depth to differing degrees in differing contexts. However, each contribution is probabilistic. Distance will *likely* cause these cues to be present, and the presence of the cues will *likely* indicate distance, but this is not always the case. For example, we can generate

---

[8] Arguably, the set of problems I am discussing here under the heading of ecological validity better matches Brunswik's notion of 'representative design' rather than the somewhat narrower issue he denoted when he coined the term 'ecological validity' (Araújo, Davids, and Passos 2007). Though it seems worth noting that the terms have shifted, I retain the usage I find in the current literature.

optical illusions of distance by creating binocular parallax using a stereoscope, or 3-D glasses in a movie.

Given this multifaceted probabilistic understanding of perception, Brunswik argues that we will never be able to experimentally control all of the environmental variables that might influence distance perception. Perhaps, were such perfect control possible, one observation might be enough to study the impact of a single variable. And perhaps also then, a precise, lawlike relation between the variable and perception might be articulable (if one exists). But such control is not possible, so no such lawlike relation should be expected.

Instead, Brunswik argues that we should sample across situations in the same way we sample across individuals. The set of situations sampled should be representative of the situations that someone encounters in real life:

> For adjustment to, or cognitive attainment of, a single stimulus variable, such as distance, one would have to secure perceptual estimates of distance in a set of situations representative *of all the situations and conditions in life* which require judgment of, or adjustment to, distance. (Brunswik 1943, 263, emphasis mine)

Thus, we require a large (perhaps huge) number of observations in various contexts in order to assess the impact of any one variable. In his own study, he accomplished this by asking people to judge the size of visible objects in their environment as they went about their otherwise normal day.

This approach has obvious similarities with my own argument that evaluation of theoretical claims requires many experiments (Chapter 1). However, I would resist Brunswik's analogy between cross-situational aggregation and cross-individual aggregation. Based on this analogy, Brunswik aggregates across situations by calculating an *average* influence, or a *weight*, for each variable that contributes to performance on the discrimination task. The result is that the contribution of each variable is reduced to a single number that summarizes its influence across conditions.

Like Mook (1983), Brunswik resists the direct generalization of a single experimental result by inserting an extra step between the result and the generalization. The difference is the step each describes. Mook inserts the evaluation or modification of a theory, which is in turn generalized, while Brunswik inserts the pooling and averaging of effects across representative situations, which is then generalized. The strategy of inserting some inferential step in the middle seems right, and each of these strategies captures something important for that step. Mook is right that it is primarily the theory (or model) that is

140 ECOLOGICAL VALIDITY

intended to be generalized (at least for claims about cognition, the main subject here). Brunswik is right that we need to look at a large number of situations to understand the implications of any single measurement or result. However, each also gets something wrong. We cannot directly evaluate a theory from a single experiment, as Mook requires. And a more modern approach to cognition requires a more articulated understanding of the reaction to various situations than the simple average that Brunswik describes.

My suggestion for how to attempt to address ecological invalidity is a combination of the two strategies. Evidence from various related/similar situations is pooled and evaluated, as described by Brunswik. However, 'pooling' is structured by theory, and theoretical evaluation comes before generalization, as described by Mook. This is, in fact, the strategy I described in Chapter 1; in this chapter we arrive at the same place even though we follow a different route.

The key part, as described in Chapter 1, was that evaluation of a particular statistical hypothesis be kept separate from evaluation of attending substantive (theoretical) hypotheses. The structured pooling of results is one way to build up from statistical results in a single experiment to a theoretical evaluation. There are a number of specific ways that it can proceed. I described two general ways one might structure the project. First is by dividing a capacity into a number of different dimensions. Second was in building an associative model to match input–output mappings (see also Chapter 3). These are not exhaustive, and whatever strategy works best is an open question. However, I note here two features of either strategy. Firstly, they compile evidence from a large number of experiments as a means to evaluate theoretical models. Secondly, the way that the evidence is compiled is not theory-neutral; the specific structure in either case will be influenced by theoretical understandings of the process, or task, involved.

Even accepting these points, we can venture tentative, informed partial generalizations of an individual result. These must be hedged and structured by what we already know. Moreover, this structuring evaluation of pooled results can help identify whether the same process is in fact behind performance in all the (supposedly related) tasks; ideally, one should see patterns that fit a reasonable characterization of the cognitive capacity. If some result stands out as a poor fit, perhaps the animal performed that task a different way. So this helps with the internal and external validity worries. And insofar as it can support tentative generalization, it might help with integrating work in the lab and field. However, I argue that there may be particular *ways* of 'pooling' or looking across experiments that are especially helpful for ecological validity. This is the substance of my positive proposal, which I turn to now.

5.7 Anchoring Experiments

When researchers discuss ways of addressing ecological validity, the goal has typically been to figure out how to build a more natural version of a particular experiment. This is a useful approach in many circumstances. However, due to all of the ways that an experiment can be artificial, there is little one can say in general. I have suggested that the idea of looking *across* a set of experiments (instead of looking at experiments one by one) provides a different kind of strategy. Specifically, looking across a set of experiments that are artificial in different ways and to different degrees can provide a more systematic understanding of how the process works, and how natural and artificial factors contribute. This can be a strategy in its own right, but it can also help the project of making individual experiments more ecologically valid by identifying which sorts of artificiality matter in a given task and/or species. In either way, I argue it can help scaffold integration of work in the lab with work in the field.

The strategy, in its prototypical form, would start with a behavior that has been observed naturally. It would then find one or more ways to approximate that behavior in a more controlled setting, or to intervene in the field setting. Either way, once we have an experimental paradigm that reasonably closely matches the natural behavior, that experiment can serve as a kind of *anchor* keeping future laboratory work connected to the natural behavior. If the animal performs under those conditions, the task, stimuli, or setting can be varied in a large number of ways. Performance on these variants is then compared back to the original task. I call this original experiment the *anchoring experiment*. As we run variant experiments, those remain anchored to a real-life task through that experiment, even if the variant experiments themselves include outlandish and artificial components.

This strategy reduces concerns about the ecological validity of any individual study, because the issue is in the *difference* between the current experiments and the anchoring experiment, and thus back to a natural behavior. The remainder of features can be held fixed (or reasonably closely to fixed, or perhaps randomized). Such a project would probe the shape of the capacity in controlled settings: what changes in task and context can the capacity accommodate, and which not? How does it respond to the changes it can accommodate? If the animal behaves more or less consistently across variants, this framework would also allow much greater confidence that the capacity in question is what is being tested because we can relate it back to a natural behavior. If the behavior breaks down or changes significantly (e.g., the animal fails a variant of a task they had previously succeeded in), then it doesn't matter whether the cause of the failure is artificial or naturalistic; either way, this shows the range of tasks to which the capacity applies (see Figure 5.2).

Work on bee foraging provides an instructive example. As mentioned, bees make a good example here in part because they, by nature, live in something of a gray area between wildness and domestication. As a first, most natural step, researchers can observe bees at the hive and at possible foraging sites in which flowers are blooming. For instance, Visscher and Seeley (1982) recorded waggle dances performed by bees in a hive returning from foraging. They decoded these dances to map the locations at which bees were foraging, and showed that they change substantially over just a few days. Researchers can study this further by introducing artificial feeding sites. Seeley, Camazine, and Sneyd (1991) introduced two artificial feeders with sugar-water, trained a small number of bees to visit each, and then tracked recruitment by counting bees arriving. In the crucial manipulation, they changed the percentage of sugar in the mixture. Within just four hours of the change, substantially more bees were visiting the higher-sugar feeder. This strategy of introducing artificial feeders is the same as in the lake experiment discussed earlier in this chapter. Research like this gives a better sense of how hives coordinate as units, as well as how individual bees use the waggle dance. For instance, bees can track which times of day a site is most productive, and even sometimes spontaneously waggle dance in the night, apparently to 'remember' productive locations (Chittka 2017).

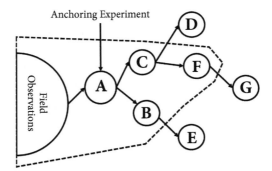

Fig. 5.2 A depiction of the anchoring experiment strategy. Researchers observe some real-world behavior and design an experiment version that is close to it (experiment A); this then serves as the anchoring experiment for further variations, and variations on variations (Experiments B–G). When the behavior breaks down in some significant way (the dashed line), we have discovered a limit of the capacity.

ANCHORING EXPERIMENTS   143

Work on the specific decisions by individual bees about which individual flowers to visit has a similar spread from naturalistic observations to artificial experiments. Any given patch of wildflowers is likely to have many kinds of flower, which may differ significantly in the nectar and pollen they provide. As such, bees will typically stick to a single species within a patch. This is apparent from observations of bees moving through a flower patch (Darwin 1877, Ribbands 1949). One can also test the pollen carried by foragers as they return to a hive: 95–99 percent of bees have only one kind of pollen (Free 1963, Seeley 2014). Researchers asking these questions can then introduce artificial flowers. Artificial flowers allow for control over a number of variables: the sensible properties of the flower, such as shape, color, and scent; the amount of reward provided by any given type of flower (generally a sugar-water 'false nectar'); and even the location of flower types relative to one another in a 'patch.' 'Scout' bees that discover a patch of two different kinds of artificial flowers will sample them and determine which are most productive (Waddington and Holden 1979). Recruits that have learned of a flower location from a waggle dance learn the scent of productive flowers by smelling the bee dancing in the hive. Using artificial scents, Koltermann (1969) found that recruit bees arrived at the right kind of flower with 99 percent accuracy. Bees learn by scent very quickly, visiting flowers with the same scent 90 percent of the time after a flower gives a reward, and 100 percent within three more rewarded visits (Seeley 2014). They can also learn by color and shape, but more slowly, requiring between five and twenty rewarded visits to get to 80 percent accuracy. Note also that scent learning, which requires multimodal integration, is among the best understood roles for the mushroom bodies (Heisenberg 1998; Strube-Bloss and Nawrot 2011), structures in the bee brain that are often thought to underlie their most sophisticated behaviors and perhaps are reasonable functional analogues of some human learning mechanisms (Mikhalevich and Powell 2020). In experimental settings, bumblebees have also been found to learn observationally, selectively visiting artificial flowers like the ones that they saw other bees visiting (even copying artificial bee demonstrators; Worden and Papaj 2005). Arguably, the bee arithmetic experiments could be seen as an experimental task that extends studies of foraging to a kind of artificial extreme. At the surface level, it involves identifying cues about where to find food. Whether this is the right way to frame it is a harder question (more below).

This is a brief summary, but the idea should be clear; researchers have built up a systematic understanding of the foraging strategies of bees based on field observations, along with experimental paradigms that bridge the field and the lab, allowing systematic variation of factors that may matter for foraging. In

144 ECOLOGICAL VALIDITY

addition to foraging itself, these studies have shed light on two of the most interesting bee behaviors for discussions of animal cognition, the waggle dance and multimodal learning of which flowers are productive. This is an ideal case of the anchoring experiment strategy I suggest here. Bees are something of an ideal subject, and perhaps it is easier to do this with bee foraging than many other behaviors by other species. But the strategy here can generalize.

In fact, I argue, something like this has long structured research programs in comparative psychology (and psychology more generally). Certain landmark studies or experimental paradigms have always become touchstones that researchers vary in all kinds of ways. Perhaps the most influential examples through the twentieth century were Pavlov's work on what came to be known as classical conditioning, and Skinner's work on operant conditioning. Each of these basic experimental setups has spawned uncountable variants applied to learning different relations, with different reinforcement programs, different rewards, in different species, and so forth (e.g., McSweeney and Murphy 2014). The false belief task has played a similar role in developmental psychology, where innumerable variants have been designed to test children at different ages in different ways (Wellman, Cross, and Watson 2001), and has recently expanded to work on great apes. Tests of causal cognition like the trap tube task have been varied in a number of ways (Visalberghi 2018). This is a natural way for experimental programs to progress as researchers probe the capacities behind performance on the original task.

Sometimes even a single study can systematically vary the task. Hoffman, Cheke, and Clayton (2016) studied the ability to pull a string and haul a reward into reach. This is something bumblebees can do, and even learn socially, by observing other bees, as with the ball-rolling task discussed above. Alem et al. (2016) even found that the behavior could be passed down through multiple 'generations' of non-trained social learners. However, they also included a variant of the task in which the string was 'coiled' in front of the sugar source, such that pulling did not immediately cause the reward to (visibly) move closer. Because the bees failed at this variant (which other species have passed; Jacobs and Osvath 2015), Alem et al. attribute the ability to "a basic toolkit of associative and motor learning processes" (2016, 1). However, a broader comparison of systematic variants could give a fuller sense of just how much an animal understands the causal structure involved. Studying scrub jays (instead of bees), Hoffman, Cheke, and Clayton (2016) did just this. They tested a variety of physical orientations and patterns in the string, including strings that cross one another and strings that did not connect to anything. They conclude: "Scrub-jays had a partial understanding of the physical principles underlying string-pulling but relied on simpler strategies such as the proximity rule to solve the tasks" (Hoffman, Cheke, and Clayton 2016, 1103). The key strategy is taking a single task as the core and systematically varying parameters of interest to observe the overall pattern.

So, there is some precedent for this kind of approach. The difference I am suggesting is that these systematic variations are attended to more deliberately and systematically at a larger scale, and (when ecological validity is a concern) that the touchstone is a natural behavior rather than an experiment that falls into this role for other reasons. There also need not be a single 'anchoring experiment' in any real case; there can be many that do different work. Moreover, the choice of anchoring experiment and how it can be varied in following experiments will require theoretical interpretation going in, and will plausibly be open to revision as work goes on. The actual structure here, I take it, would vary from case to case. However, as a general framework for thinking about ecological validity and artificiality in laboratory experiments, it provides a valuable structure which can help address the resulting worries.

Most obviously, I think, the anchoring experiment approach helps with the goal of integrating the lab and field. In ideal cases, it can build a direct bridge between them, as has been done in the bee foraging literature just described (in that literature, experiments are often performed by ecologists in biology departments). It can help address worries about external validity and generalizability to the wild, because the network of related experiments helps trace connections to wild behaviors. In the best case, each experiment tests how that wild behavior *would* change in response to the task variations under study. Finally, it can help address worries about internal validity, because the ability to trace performance back through related experiments and wild behaviors can scaffold arguments that the same capacity is driving behavior in all of them. None of these is guaranteed, things can go wrong in the experiments, and we can be wrong about the connections between them. None of them can completely *eliminate* the worry. Even so, they can scaffold greater confidence in the interpretation of experimental results.

I also think this strategy can help interpretation in light of the artificial lives of laboratory animals. If performance on the anchoring experimental task is close enough to natural behavior, we have reason to think the life history of the lab animals does not matter in the task (of course, it may become relevant in one of the later variants in ways it was not in the original, but we may just need to live with that risk). We might sometimes be able to go one step further: an anchoring experiment could actually include animals in the wild. Bees are not an obvious answer here, since even bees in the lake experiment were kept in

146   ECOLOGICAL VALIDITY

a hive. However, there is a tradition of experimental intervention with wild populations. As a famous example, Cheney and Seyfarth (1980) played recordings of various calls from members of a troop of vervets over speakers. Similar studies have been done with elephants (O'Connell-Rodwell and Wood 2007) and marine mammals such as whales (Deeke 2006). It will not always be possible (or ethically preferable) to intervene on wild populations in this way, but where it is, the strategy can help.

Anchoring experiments could structure the large-scale design of a research project and the specific experiments performed within it. This part of the claim may look frustratingly utopian to experimentalists; actually doing this properly would require a *lot* of experiments. Fair enough. I do not argue, though, that this is always necessary. I also think that a reasonable approximation can be accomplished in some research areas by reinterpreting existing research. I already did this, in effect, in the way that I set up the bee examples. In other cases where we have research in the field and the lab, we might be able to do the same. In effect, we can find anchoring experiments in the existing literature by reinterpreting patterns of experiments that have already been done rather than by starting from scratch with a new research program. Timberlake (2002) argues for an approach to integrating laboratory and field studies that involves just this kind of reinterpretation of existing experiments. He even targets classical behaviorist paradigms of operant conditioning and maze learning. In the process of tuning those paradigms to get results, he argues, the researchers (perhaps inadvertently) built them to match, and recruit, niche-specific capabilities of the animals. As minimal examples, operant conditioning experiments with rats require the subject to press a lever to be rewarded, while experiments with pigeons require pecking. This makes sense because rats naturally dig and manipulate with their paws in gathering food, while pigeons peck. Similarly, he argues, experiments that place rats in mazes recruit tendencies to follow established trails through the underbrush and as burrows, which are significant to their lives in the wild. Plausibly, there is some similar connection between lab and wild behaviors for most popular experimental paradigms. These connections can provide a strong start to an anchoring experiments approach, without having to do any additional experiments.

We could also do more experiments to backfill the gap between some existing research and more natural tasks. I suggested, for example, that the bee arithmetic experiment could be seen as an extreme variant of bee foraging tasks. One could test whether there is anything to this by systematically experimenting with task variants that share some features of the arithmetic experiment but still look more natural, building a connection between the two.

For example, it's an interesting difference between the arithmetic and normal foraging that the cues for which 'flowers' will produce are separated from the flowers themselves, and that the two cues only make sense in relation to one another (one at the entry to the maze, and one labeling each of the options). Future experiments could build up understanding of how bees interpret these types of cues compared to cues on the flowers themselves. Indeed, I think this would help address the interpretive questions about that experiment because it would give better context for how the bees interpret the task. If we could build similar bridges between the wild and the networks of experiments varying a touchstone experiment, as just mentioned, we might be able to connect lots of work more systematically to natural behavior. The point, either way, is that the anchoring experiment strategy need not imply a wholesale revision of experimental design strategies, and we can at least approximate it with much of what we have.

The overall goal is to be able to evaluate patterns of response across a set of contexts, which ideally include both natural behaviors and systematically varied experimental designs. The closer we can come to the ideal in any given research program, I argue, the better shape we are in to avoid the challenging implications of unnaturalness. This can be a way of actually building research programs, or, more abstractly, it can be a way of thinking about those programs and interpreting experiments within them.

5.8 Conclusion

Ecological validity can be taken seriously as an issue without simply being treated as a reason to dismiss or shut down research. Building research programs around anchoring experiments is not an easy solution, and I recognize that it would require reshaping much of the field. However, I think there is a case to be made that as comparative psychology and cognitive ethology come closer together, we are already moving in something like this direction. Less ambitiously, though, perhaps thinking about experimental design and interpretation in these terms can help. Different research projects using different species may implement this in different ways, but the core idea can remain even with substantial variation in the specifics.

There are three main practical problems that artificiality might introduce which have been described in the chapter: internal invalidity, external invalidity, and difficulties with integration. This approach alleviates (though perhaps doesn't eliminate) all three: we can have some more confidence that we are testing the intended capacity, and perhaps more quickly identify when we are not (internal validity); we can make better informed generalizations (external validity); we can build frameworks to help integrate work in the lab with work in the field (integration). Moreover, we can understand the influence of artificial aspects of the experiments themselves, as well as the unusual background and training of laboratory animals.

The project described here is difficult, and my suggestion that it will be fruitful in practice is admittedly speculative. Even so, I think this is the best approach we have to worries about ecological (in)validity. If we can anchor the behaviors in a set of experimental tasks to natural behaviors, we stand a better chance of understanding the implications of that experimental behavior for life in the wild.

6

Sample Size and Generalizability

6.1 Introduction

If you flip back through this book and track down the empirical papers I mention, you'll find reports of experiments that include very few individual animals: some include a few dozen, but many include only a handful of animals and, in one case, only one. This is broadly representative of comparative psychology. Experimental studies in comparative psychology are often severely restricted in the number of animals that they can include. This limitation leads to questions about whether results should be generalized to other members of the species. Coupled with characteristically high variance and small effect sizes, it has also recently led to some concern about replicability in comparative psychology. The ongoing 'replication crisis' in other parts of psychology has made this especially pressing. I focus specifically on the question of whether experimental results with small samples can be generalized to claims about the species overall, or any other taxonomic group (as opposed to the last chapter, which was interested in whether we can generalize experimental results across situations).

Issues with sample size have only recently been explicitly confronted in the comparative psychology literature (e.g., Stevens 2017; and an extended discussion in the May 2021 edition of *Animal Behavior and Cognition*), and have not yet been seriously addressed in the philosophy literature on animal minds. Historically, the worry has only come up in a piecemeal way, authors pointing to sample size as a way of criticizing a particular experiment or lamenting sample size as an ineliminable limitation of their study (in this regard, discussions of sample size have been very much like discussions of ecological validity, as discussed in the last chapter). Rather than make predictions about whether studies would replicate, or advocate changes to experimental practice that might allow increased sample sizes, my focus is on how to interpret studies with the samples we have. As with the other challenges, I think small sample size changes the way we should interpret studies, but it doesn't undermine the work. Small sample work is still valuable.

Seven Challenges for the Science of Animal Minds. Mike Dacey, Oxford University Press. © Mike Dacey 2025.
DOI: 10.1093/9780198928102.003.0007

Replicability and generalizability are tightly linked questions; it might be tempting to think of the question of whether a result generalizes as being exactly the same as the question of whether it would replicate. This is not quite the case though: experiments in comparative psychology often use within-subjects experimental design. In such a study, the statistical power may be sufficient, but the relationship between generalizability and replication will depend on which parameters one intends to resample in the replication (Machery 2020; e.g., do we resample from the population, or do we use the same individuals at another time?). As such, I see the issues as separable at least in principle (more to come), and I focus on what I take to be the underserved question of interpretation and generalizability of small sample studies.

My discussion of this issue is not intended as a criticism of the field or its experimental and statistical practices (though I will take issue with some interpretations). As I articulate below, there are intrinsic features of the field of study that drive these limitations, so, as with the other chapters, this challenge arises because of the nature of the subject of study. I also take it that most researchers in the field are aware of worries about sample size. These researchers may disagree on just how much of a problem it is and what to do about it. There is plenty of room to reasonably disagree on these questions. My goal in this chapter is to provide one way of thinking about small sample size research that faces up to its limitations without dismissing it outright. I noted in the introduction that it can be difficult to frame a discussion around challenges for the science while keeping my positive orientation at the front. I am most sensitive to this difficulty in this chapter. Even where I must discuss limited samples at length (e.g., in Section 6.3), the payoff will be a positive view.

This positive view includes a significant shift in the interpretation of and evidential weight granted to studies with very small samples. These studies share some features of experiments as standardly viewed but also some of the features of other sources of evidence, such as case studies and anecdotes. Before we can get to the details, though, we have to get a sense of the sample size limitations themselves, and their significance. For a sense of their significance, I turn to the history.

6.2 Animal Anecdotes and the Founding of Comparative Psychology

Comparative psychology arose as a field around the end of the nineteenth century, as experimental methods emerged in psychology more generally. The scientific backdrop, at the time, was shaped by debates over Darwin's work on evolution, at that point only a couple of decades old. Darwin's views on evolution suggest that continuity across species is the rule, because evolution works with small changes that build up slowly over many generations. Nowhere was this result more significant than when it came to the mind. The fiercely held conventional wisdom at the time was that human minds were truly different and special.

Darwin supported the general argument for continuity in the mind with (what he took to be) more direct evidence. Much of this evidence was based on anecdotes about clever or otherwise commendable behavior in animals. For instance:

> I will give only one other instance of sympathetic and heroic conduct in a little American monkey. Several years ago a keeper at the Zoological Gardens, showed me some deep and scarcely healed wounds on the nape of his neck, inflicted on him while kneeling on the floor by a fierce baboon. The little American monkey, who was a warm friend of this keeper, lived in the same large compartment, and was dreadfully afraid of the big baboon. Nevertheless, as soon as he saw his friend the keeper in peril, he rushed to the rescue .... (Darwin 1871, 75)

This anecdotal approach continued in the work of George Romanes, Darwin's appointed successor on psychological topics. Describing similar animal heroism, Romanes says (also reporting the story secondhand) that a column of ants "rushed to the rescue" of an individual pinned down by a rock, and "[t]his observation seems unequivocal as proving fellow-feeling and sympathy, so far as we can trace any analogy between the emotions of the higher animals and those of insects" (1883, 48–9).

Around the turn of the twentieth century, authors such as C. Lloyd Morgan (1894) and Edward Thorndike (1911) rebuked this reliance on anecdotes. To be a science on firm founding, they felt, the field would need to reject anecdotal methods in favor of more rigorous experimental methods. The resulting shift, so the common foundation story goes, brought comparative psychology into its own as a rigorous science (e.g., Galef 1996; Shettleworth 2012).

It is easy to see what is objectionable about the way Darwin and Romanes use anecdotes. They relay their stories secondhand without scrutiny and leap to a heroic interpretation without considering other explanations. There is also a particular worry that work on animal minds will be systematically biased by the unconscious human tendency to anthropomorphize; to interpret

152 SAMPLE SIZE AND GENERALIZABILITY

animal actions in the same ways they would interpret human actions (see Chapter 2; Dacey 2017a). Narrative anecdotes seem particularly ripe for such a bias. They often presume intentions behind the action, and often elicit emotional reactions and bonds with characters that may threaten impartial scientific analysis. To put it bluntly, replacing anecdotes with experiments makes comparative psychology look more like other successful sciences (Thorndike 1911).

There are many reasons that scientists in general reject anecdotes. I see the key concerns about anecdotes as follows (listed to aid later discussion; Section 6.6). These concerns overlap, and are not exhaustive, but aim to capture the main thrust:

1. Anecdotes can be cherry-picked to make a predetermined point.
2. We lack control over, and knowledge of, background conditions of anecdotal events.
3. Anecdotes are narrative in structure rather than providing analyzable data.
4. Anecdotes are non-repeatable (non-replicable), and so can't be confirmed independently.
5. Anecdotes don't support generalization.

Controlled experiments do not face these concerns. One cannot pick and choose which individual responses in any given experiment make it into the data (though one can choose which *experiments* to report, which will be discussed below). A good experiment is defined by control over the variables that can influence behavior, manipulating variables of interest, while holding others fixed or washing them out in averages. Experiments produce evidence in the form of data, which is cold, dispassionate, and suited to statistical analysis. As a result, when done well, experiments are replicable (worries are noted in Section 6.4), and they can support generalization.

Summing up, a common foundation story for comparative psychology tells that it came into its own as a science when it chose experiments over anecdotes. In this framing of the field, anecdotes are opposed to, and mutually exclusive of, experiments. As is often the case with the histories that scientists tell (e.g., Kuhn 1962), this functions both as a (simplified) description of events and as a normative reinforcement of the importance of experimental practices for comparative psychology. However, when we dig into the reality of experimental practice in comparative psychology, this story might look more like an unattained aspiration.

6.3 Sample Sizes in Animal Labs

There are a number of practical reasons that sample sizes are limited in laboratory behavioral experiments with animals. Animals must be kept and cared for, which is expensive, complicated, and time-consuming. As a result, there are usually a limited number of animals on-site as candidates for any experiment. Ethical concerns often dictate that the number of animals involved should be as low as possible—both the number in specific experiments and the number kept in captivity. Individual experiments usually require time-consuming training for all participants, so not all animals in a given facility can participate in each study. Similarly, experiments sometimes require that animals be 'naïve' to the tasks at hand, which rules out animals that participated in earlier studies. This may also require that researchers strategically leave some naïve for future work. There are also often basic tasks that an animal must successfully perform to even participate in the experiment, and those who fail will be excluded. Putting these together, the implications are stark. Experiments frequently include samples of individual animals in single digits, and sometimes only one or two.

As an illustrative example, I will focus on Inoue and Matsuzawa's (2007) paper "Working memory of numerals in chimpanzees." This paper compares human and chimpanzee performance on a short-term memory task. The authors state their conclusions unequivocally: "Our study shows that young chimpanzees have an extraordinary working memory capability for numerical recollection better than that of human adults" (1005). The paper became something of a media darling and has been cited extensively (often uncritically) in the academic literature.

The task is as follows. Participants (human and chimpanzee alike) sit in front of a computer screen. The screen quickly flashes several digits in random locations on the screen (all shown simultaneously). After a presentation of a few hundred milliseconds (650, 430, and 210 ms in different trials), each digit is masked with a white square. Participants are asked to then tap the mask squares in the order of the digits previously at each location. The researchers measure both response time and accuracy. The task is meant to test the ability to rapidly store working memories of the visual scene (210 ms is too fast to saccade through the sequence).

Inoue and Matsuzawa began with fourteen chimpanzees on-site. Of these, they involved six chimpanzees in the task (three mother–child pairs). While all six were able to learn the most basic task, only four could manage a sequence of five numerals, which was the number used in the key test. So, following the

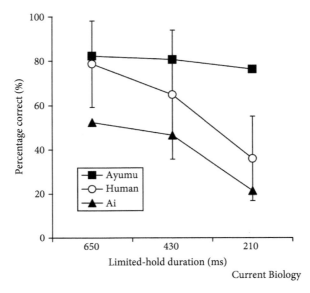

Fig. 6.1a Data from Inoue and Matsuzawa (2007). The accuracy of two chimpanzees, Ayumu and Ai, is compared to a human average at each presentation duration. This and Fig. 6.1b reprinted from Sana Inoue and Tetsuro Matsuzawa, "Working memory of numerals in chimpanzees," R1004–5, *Current Biology*, 17(23), © 2007, with permission from Elsevier.

progression of limitations described at the opening of the section, we move from fourteen to six to four animals. The actual data presented in the paper, however, compares individual chimpanzees, one at a time, against a human average (human n = 9 in one experiment, n = 12 in another). So for each actual comparison, chimpanzee *n* = 1. In fact, the assertion that chimpanzees perform better than humans seems to be based on a single chimpanzee—Ayumu, the best chimpanzee performer (see Figure 6.1a). Based on the data presented in supplemental materials (see Figure 6.1b), all three of the younger chimpanzees, including Ayumu, performed *faster* than the humans, but only Ayumu performed with higher accuracy than the human average. So, the key claim here seems to be based on a simple size of one.[1]

This is one experiment in one paper, but the core worry generalizes; I am not basing my argument on a single anecdote! Figure 6.2 shows the number

---

[1] Even if we take the assertion of 'better' performance to simply mean 'faster,' it is based on three individuals. While it is not my focus here, I will point out that the majority of human participants performed with better accuracy than Ayumu, even though his performance beat their average.

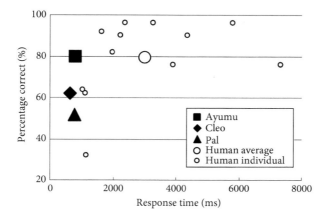

Fig. 6.1b Data from Inoue and Matsuzawa (2007), showing response time and accuracy with 650 ms presentation of stimuli before mask. This is the same condition as the leftmost data points of Figure 6.1a (supplemental materials, figure S2—the paper does not provide comparable data for other durations).

of individual animals included in every experiment published in 2019 in four of the top journals that specialize in animal cognition: *Animal Behavior and Cognition, Animal Cognition, Journal of Comparative Psychology,* and *Journal of Experimental Psychology: Animal Learning and Cognition.* We recorded the number of individual animals used in each *experiment*, and not the number per paper or per experimental condition.[2] Out of 151 experiments reported in 90 papers, 98 experiments include data from 20 or fewer than 20 animals, 50 from 10 or fewer, and 20 from 5 or fewer. Wherever, exactly, we place the boundary for small sample size, small sample studies are published regularly.

I present this data with an important caveat: it is not meant to present a statistically rigorous picture of the field. The goal is simply to demonstrate that small sample studies are common, not that they dominate or make up any particular percentage of research in comparative psychology. This is why I am not presenting any summary statistics. These would risk being misleading and are not the point anyway. Moreover, the selected set of journals cannot be taken to provide a full cross section of the field. These journals are among the top that focus on animal cognition. They were chosen in advance, along with the date range, to limit potentially subjective decisions to include or exclude

---

[2] Were we concerned with statistics, the number per experimental condition would have been appropriate. In general, looking at conditions would have reduced the observed sample sizes, but given the diversity of experimental designs, it is not always easy to divide up conditions. More below.

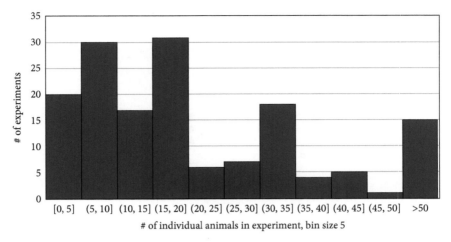

Fig. 6.2 A histogram of all experiments published in the journals *Animal Behavior and Cognition*, *Animal Cognition*, *Journal of Comparative Psychology*, and *Journal of Experimental Psychology: Animal Learning and Cognition* in the year 2019, sorted by the number of individual animals in reported data. This includes 151 experiments described in 90 papers—any experiment that made an intervention, laboratory or field; 54 papers were excluded, as they did not present new behavioral data (29 papers), were unable to report the number of animals involved (7 papers), were purely observational field studies (6 papers), or used only human subjects (12 papers). Thanks to Abe Brownell for collecting this data.

particular studies.[3] However, they are not the only such journals, and animal cognition studies are often published in more generalist journals as well. Some of these, including *Nature* and *Current Biology*, (the publication venue of Inoue and Matsuzawa [2007][4]), are generally considered the most prestigious venues. Finally, work involving different species tends to face the limitations discussed above to differing degrees, and so tend to face different pressures on sample size.[5]

Lastly, this data should not be taken to reflect the statistical power of the studies or their probability of replication. I assume for present purposes that the statistical methods applied by these researchers are sound. The aims of

---

[3] For instance, having to decide which studies in a more generalist journal would count as a behavioral study of cognition as opposed to neural/anatomical, or something else. Note: we had intended to include the journal *Animal Behaviour*, but the coronavirus pandemic cut our study short.

[4] The sample size for that experiment would have been four in this analysis.

[5] For example, with all caveats just presented in place, we did find in our sample that work with mice and rats appeared less likely to have very small samples.

these papers are multifarious, but many use within-subject designs or make many observations from the limited sample of participants. The analysis here focuses on the number of individuals in an experiment simply because it is a straightforward number to collect across experimental designs, and an easy one to understand. Moreover, I think it is actually *more* useful for thinking about whether a conclusion generalizes than replication itself. For an example of what I mean, let's return to the Inoue and Matsuzawa (2007) study. I have every confidence that if a different group of researchers (in a different lab, say) attempted to replicate the study *with Ayumu*, Ayumu would replicate the result.[6] It also seems plausible that if researchers ran the same experiment using several chimpanzees from a different population, they might find one who can match Ayumu's performance. At the same time, we cannot conclude that this level of performance is a trait of chimpanzees in general. Most likely, whatever individuals one might find that perform at that level are exceptional. There may still be some things we can learn about chimpanzees in general from that conclusion. But the blunt assertion that "chimpanzees outperform humans," based on the observation that one chimpanzee does, at best elides interesting questions about what we can actually say about chimpanzees in general.[7] These are the questions I am interested in.

Summing up, for practical reasons experiments in comparative psychology are often severely limited in the number of individual animals they can include. Many studies are done with samples in single digits, and studies with only one or two individuals are not uncommon. With these sample limitations in mind, especially in the limit case of $n = 1$, it seems reasonable to ask: just how much has the field progressed beyond its anecdotal roots? Are these not, in an important sense, merely anecdotes that take place in a laboratory setting?

6.4 The Replication Crisis and Comparative Psychology

To the extent that these issues have arisen in the literature, it has come up in the context of questions about replicability. Even though I take replicability and

---

[6] Borrowing from my characterization in Chapter 1: statistical methods show that "there is an effect here." So within-subject design on a single individual can provide good evidence that that individual's performance is real, and not merely a fluke of statistical noise. It's another question what this means for the species more generally.

[7] This is compounded by the fact that Ayumu here is an outlier among even the top performers: only those individuals able to perform the basic task were included, and Ayumu's performance was an outlier among them. There are also concerns that the life history of laboratory animals makes them unrepresentative, which I set aside here (see Chapter 5).

generalizability to be separable, they are related. As such, these discussions will be helpful. In recent years, other branches of psychology have instituted reforms to address prominent and repeated replication failures (Romero 2019). Despite the obvious worry that small sample sizes leave comparative psychology vulnerable to these same problems, the field has only just begun to respond (Beran 2018; Farrar, Boekle, and Clayton 2020).

In perhaps the first paper to meet the issue head-on, Stevens (2017) argues that there are several reasons to think that comparative psychology is vulnerable to replication failures (though, as he notes, the regular use of within-subject design protects against many statistical anomalies). He makes several recommendations for the field to address these concerns. Researchers have begun to implement some of these recommendations. I will mention some of his recommendations which inform the current discussion. One is for researchers to preregister their methods before the test, or alternatively for journals to adopt the practice of registered reports, in which a journal accepts or rejects a paper based on methods alone, before experiments are run. These practices have grown recently in social psychology. The purpose is to prevent fishing-expedition approaches to studies and statistical analyses. These can lead to cherry-picking which experiments are reported, and "P-hacking" by, for instance, simply trying every statistical analysis one can think of until some analysis returns a significant result. In 2018, the journal *Animal Behavior and Cognition* began accepting registered reports (Vonk and Kraus 2018). The editors indicate that initial uptake was slow (Beran 2020) but is perhaps picking up (Vonk and Beran 2022).

For reasons discussed in the last section, worries about sample size pose a deeper problem for comparative psychology than most branches of psychology. For instance, in response to replication worries, social psychology has massively increased experimental sample sizes, using online platforms for data gathering. Comparative psychology has no such option. Nor can it entirely avoid the limitations discussed above. Nonetheless, Stevens makes some recommendations that could help increase sample sizes. First, different labs can collaborate and combine their subject pool. In fact, the ManyPrimates Project was launched in 2019 to facilitate collaboration across labs spanning the globe, allowing for larger and more diverse samples in studies of primate cognition (ManyPrimates project members 2019). ManyBirds (Lambert et al. 2022) and ManyDogs (Alberghina et al. 2023) have followed suit. Second, Stevens suggests that researchers can take advantage of facilities like zoos that may have larger numbers of animals available. Third, researchers can reconsider their choice of species, either by running studies pooling multiple species or by switching to species that are easily available in the community, such as dogs.

In another discussion, Farrar, Voudouris, and Clayton (2021) make three main suggestions of their own. First, they suggest sampling a larger and more diverse population not only of individuals but also of situations, materials, sites, times of day, measurements, and so forth—as much as is feasible. This suggestion closely aligns with Brunswik's (1943) approach to ecological validity discussed in the last chapter, and thus with my own suggestions there. However, I would not require that a single experiment sample so broadly; this can occur across experiments/papers as well. In the same volume, though not in direct response to Farrar, Voudouris, and Clayton (2021), Halina (2021) notes that, in the animal mind-reading debate, sets of different experiments have been done which resample various of these features, while keeping others constant. She characterizes these as replications of an original experiment because they test the same (substantive) hypothesis. I am not inclined to count these as 'replications' and instead view them as independent sources of converging evidence. But it seems worth noting that both of these papers are getting at the same idea of looking across large volumes of data with systematic variations, either within a single experiment, (incomplete) replications, or across sets of experiments. The main value each contributes towards generalizability is substantively similar. Farrar, Voudouris, and Clayton's second suggestion is that, where it is impossible to systematically vary these factors within an experiment, researchers should move in the opposite direction and exercise exacting control over detail. A better-controlled experiment is more likely to give the same results in different populations. Thirdly, and finally, they suggest that researchers model variance or, failing that, artificially inflate the uncertainty in their statistical estimates in order to raise the bar needed to reach statistical significance. I think here, again, my general strategy of limiting the evidential weight applied to a single experiment and looking instead *across* experiments provides a better answer: we need not meddle with statistics if we limit the statistical result to the statistical hypotheses and recognize that generalizing or making inferences to substantive hypotheses will require at least some qualitative inference (Chapter 1; Dacey 2023).

For the most part, though, these suggestions, between Stevens (2017) and Farrar, Voudouris, and Clayton (2021), seem reasonable to me for helping to avoid replication failures. I remain most interested, though, in generalizability over replicability itself. As noted above, a study, especially with within-subject design, might replicate without providing a finding that can be generalized to other members of the species. This also cuts in the opposite direction: it

160 SAMPLE SIZE AND GENERALIZABILITY

can be hard to know what a replication failure means for generalizability. For example, Boyle (2021) notes that a failed attempt to replicate faces all of the same interpretive challenges that any other experiment does—the same challenges that have structured this book. A replication attempt may fail for various reasons, so it may be that a study that fails to replicate can still offer information worth generalizing.

With my focus on generalization, rather than replication per se, a few of the ideas raised in the burgeoning discussion of replication in comparative psychology are most interesting. First, the recommendations aimed at increasing sample size certainly help both for replicability and generalization. They are unlikely to completely address the problem across the entire field (they simply cannot operate at the scale that online data collection makes possible for social psychology), and different changes may matter more for specific purposes, but the efforts they represent are worthwhile. Second, the emphasis on large sets of data that systematically vary different parameters of experimental design also seems important—whether these come as larger experiments, sets of experiments, or replication attempts. Finally, for reasons that will become clear below (Section 6.6), I take the suggestion of implementing registered reports to be important. If papers are evaluated based on methods rather than results, it will significantly impact our interpretation of studies with small sample sizes.

6.5 Candidate Analogues

Even if the field makes these proposed changes, interpretive challenges for small sample studies remain. Researchers are not likely to completely address sample size worries in comparative psychology. But even if they do, we should still consider how to interpret small sample studies that already exist.[8] This, I suggest, requires a shift in the way we think about these studies and the evidence that they can (and cannot) provide.

To help inform this shift, I present four methods in related fields that make use of extremely small samples, even samples of one: Incident reports in cognitive ethology, lesion studies in cognitive neuropsychology, small-N design in vision science, and case studies in social science generally. All of these are generally accepted as methods in their respective fields, though the details of what they can do, and how, may be controversial. None of these is a perfect

---

[8] If there is, in fact, a role for small sample studies, then perhaps some (certainly not all) of the urgency of universally increasing sample size is removed.

analogue, but each can provide some helpful suggestions, as well as warnings, for how we ought to think about anecdotal experiments.

6.5.1 Incident Reports in Cognitive Ethology

Researchers in cognitive ethology sometimes report, and even publish, anecdotes, or 'incident reports,' of particular behaviors observed in the field. The practice is, of course, not uncontroversial (e.g., Mitchell, Thompson, and Miles 1997). Most importantly for present purposes, any particular observed behavior may be a fluke or one-off. In general, data based on repeated observation or experimentation is preferred when it is available.

To be clear, there are norms in the observation and reporting of incident reports, so second-hand, heavily interpreted, and uncritical reports of the sort we saw from Darwin and Romanes above would not fly. These more careful techniques can reduce worries about anecdotes above but cannot eliminate them. For instance, more careful reporting can avoid narrative structures that implicitly attribute intention to the animals and can do a somewhat better job at identifying possible causal factors contributing to the behavior.

Arguably, incident reports played a much more significant role historically than they do today, as knowledge, observational tools, techniques, and analysis methods have improved. Ramsay and Teichroeb (2019) report that anecdotes are included in less than 10 percent of papers published in top primatology journals, and the frequency has decreased since the year 2000. Interestingly, though, it may be that norms against publication of incident reports have exacerbated their limitations. Both Andrews (2020) and Ramsay and Teichroeb (2019) suggest developing new venues for the publication or sharing of incident reports among researchers. Such a systematic way of sharing potentially interesting observations could put researchers in a better position to identify which observed behaviors are merely flukes and which are not.

There are a few roles that incident reports play. Perhaps least controversially, they can simply identify behaviors that are worthy of future study (Silverman 1997). Along similar lines, they can help generate hypotheses for test, and can motivate the choice to pursue one research project over other candidates. The harder, and for my purposes more interesting, question is whether incident reports can provide *evidence* about cognitive mechanisms. Andrews (2020) argues that the limitations and biases of field observation complement the limitations and biases of laboratory experimentation. Most broadly, field observations don't face concerns about ecological validity (Chapter 5).

162 SAMPLE SIZE AND GENERALIZABILITY

There might be some advantages to narratively structured anecdotes as well. Narratives can often supply richer context about the individual behaving and its context than data (Mitchell 1997).[9] The fact that members of a species display some activity in the wild may provide support for certain interpretations of laboratory experiments as well: they might provide evidence of a role for a certain kind of reasoning capacity to play, or might show an experimental task to be a better, or worse, approximation to a real-life setting.

In general, incident reports seem especially valuable for understanding low-frequency behaviors, and behaviors that would be otherwise difficult to elicit in a laboratory setting. For example, behaviors that take place in dense bush, inside burrows, or underwater can be very difficult to observe in the field for practical reasons. Similarly, behaviors might be comparatively rare but still significant to the animals involved—such as mating or giving birth.

6.5.2 Lesion Studies in Cognitive Neuropsychology

Lesion studies in cognitive neuroscience present the second candidate analogue. In many of these studies, researchers test a single patient with known brain damage on a battery of tasks aimed at delimiting a certain cognitive capacity. Studies like this generally focus on two kinds of question. The first are questions about the neural underpinnings of a particular cognitive capacity. On the assumption that there is some degree of functional localization in the brain, we can identify candidate regions for some capacity when specific deficits co-occur with localized damage. The second are the so-called dissociations of capacities that might otherwise be thought to be expressions of a single system. For instance, if a deficit in experiential memory does not also bring with it a deficit in memory for facts, then we have reason to believe that the two are separate capacities subserved by separate systems (that is, episodic and semantic memory). Crucially, this includes the conclusion that the intact capacity does not *require* the damaged capacity. In other words, a functional episodic memory is not *necessary for* a functional semantic memory (necessity is a high bar, of course, so showing it fails to hold is not always showing much).

---

[9] Mitchell advocates specifically for anthropomorphic anecdotes as a way to conceptualize behavior. I am wary here; while anthropomorphism and anecdotes do seem connected, one can allow roles for anecdotes while encouraging wariness of anthropomorphism, which is the broad position I take (see also Chapter 2).

The evidential value of lesion studies has long been controversial (e.g., Caramazza 1986; Bub and Bub 1988; McCloskey and Caramazza 1988; Glymour 1994; Shallice 2015). Sample size is far from the only methodological concern that has been lodged against them, but it is the most important for present purposes. Individual patients, such as Patient K.C. for example, have participated in studies over decades (Rosenbaum et al. 2005). Caramazza (1986) argues that reliance on a single case is actually crucial for these studies. Individual deficits with damage vary so much that effects would likely wash out in any cohort study, leaving them impossible to interpret. He does recognize worries about replicability, but isn't bothered by them. The argument is that, in short, we can corroborate findings with other evidence, including similar future lesion cases if and when they come.

In recent decades, imaging and modeling techniques have gotten much more sophisticated. As a result, from (roughly) the 1990s onward, single-case lesion studies have declined in prominence in cognitive neuroscience. Even so, these studies did play a major role in several literatures, including work on memory (Shallice 2019), perception, and language.

There are several points we can take away from single-case lesion studies for considering in small sample animal studies. Firstly, it is worth repeating that in these examples a single case *can* provide evidence. That evidence may be limited, and must be considered alongside evidence from other sources, but this can still be valuable. It seems especially relevant that lesion studies played their most significant role when knowledge was limited and alternative methods were lacking. Secondly, there is even an arguable advantage to having a single case, as it allows a detailed understanding of the specific deficits, when details might be lost in a larger sample. Finally, the strongest form of evidence comes for dissociations, where one can show that one cognitive capacity is not necessary for another to operate.

6.5.3 Small-N Design in Vision Science

Smith and Little (2018) provide a spirited defense of 'small-N' studies in psychology. They do not merely argue that small-N is a manageable impediment; they argue that psychology ought to move away from large-N design towards small-N design if it wants to be a mature quantitative science and avoid replication problems (precisely the opposite of most reactions). Their primary touchstone is their own area of vision science. Small-N design, as they describe it, treats the individual participant as the replication unit. They note as a strength

164 SAMPLE SIZE AND GENERALIZABILITY

of small-N designs that some effects that can be seen in individuals might wash out in averages. For example, if individuals display a stepwise learning function, that pattern will be lost in aggregate, appearing as a linear slope.

They argue that small-N design allows more specific measurements because they can test mathematical models (that is, ratio predictions rather than mere ordinal predictions such as "performance will be lower in condition C"). These models predict a specific pattern of effects across a number of conditions. By repeatedly testing the same individuals in all of these conditions, one can achieve high statistical power. One can even move to a test that compares the model's predictions to the entire pattern of results, including effect size, across conditions (so-called model fitting). This much seems like a reasonable aspiration for comparative psychology (see Chapter 3 on modeling), but it may not be quite so simple.

Smith and Little (2018) do recognize that vision science typically has especially low variance. They attribute this to two main factors: first is the use of practiced, experienced observers as participants, and second is the manipulation of the stimulus (e.g., stimulus contrast) so that all participants operate at (or near) the same performance baseline. It's an open question how far this general framework can go in comparative psychology.[10] There are some areas where it is clearer; for example, work on associative learning models (Rescorla and Wagner 1972) and temporal discounting functions in pigeons fit models to curves of data produced by varying the time of delay (Green and Myerson 2004). Even so, to the extent that vision science represents a successful case for small-N design, some research programs in comparative psychology might benefit.

6.5.4 Case Studies in Medicine and Social Science

Case studies have long been a popular method in medicine and clinical psychology, as well as in social sciences such as political science, sociology, and economics. The idea is that a researcher focuses on a single case, perhaps an individual patient, nation, business, village, or historical event. They gather information about this case, collecting historical records, interviewing or observing individuals, visiting communities, and so forth. The goal would

---

[10] Smith and Little do note work on animal learning, going back to Pavlov, Thorndike, and Skinner, as an exemplar of small-N design, but say nothing specific to comparative psychology about how it can work; we are left to extrapolate from what they do say. For instance, they are dismissive of "is there a difference between condition A and B?" type experiments that arguably pervade comparative psychology.

be to construct a picture (usually, but not necessarily, a narrative) of that case which identifies plausible factors and mechanisms that explain whatever features of the case are of greatest interest. Forrester (1996) influentially argues that case-based reasoning is a fundamentally different kind of scientific reasoning than the types of reasoning most often treated by philosophers of science (see Millard and Callard 2020). More recently, Morgan (2012, 2020) has worked to characterize ways that researchers reason *from* one case *to* another. She is circumspect about calling this 'generalizing' because the resulting claims would not be automatic or universal (2020, 206). Case studies are explicitly not intended to generalize *all* or even *most* of their content, but, nonetheless, they are often used to draw more general lessons. So researchers are, and should be, very careful about which parts of a case are taken to apply more generally, and just how broadly they apply.

It is difficult to say much in general about case studies, as the norms and methods attached can vary significantly in different disciplines and differing subject matter. Even so, a few roles and strategies for case studies in the production of general knowledge can be identified. A role for cases in *discovering* hypotheses is (comparatively) straightforward. A good case study will often identify several aspects of the case that might play a causal/explanatory role in producing the phenomenon of interest. Thus, case studies can be fertile ground from which one can produce a number of candidate hypotheses implicating different aspects (and different combinations of aspects). Campbell (1975), described in Morgan 2012, argues that cases can 'infirm' or weaken a hypothesis, even when they cannot 'confirm' or strengthen any particular hypothesis. The thinking is that a case study might admit of several plausible interpretations that emphasize different aspects of the case as explanatory factors. Each of these interpretations might *support* different hypotheses, so there can be no univocal confirmatory role for the case. However, there will likely be possible hypotheses that fit *none of* the plausible interpretations of the case, and these hypotheses will be infirmed.

On the possibility of confirming hypotheses, Morgan (2020) suggests a couple of ways it might go. When one has multiple cases, she distinguishes between strategies that look at strings of cases and strategies that are based on comparison. As an example of a string of cases, she discusses AIDS in the 1980s (Ankeny 2011). The process of learning about AIDS involved a string of individual cases that were collectively brought under the shared disease category by noticing and negotiating which characteristics seemed to correlate in the ways relevant to a shared disease. The process of identifying the actually predictive characteristics and symptoms involves generating a series of tentative generalizations which are checked as new cases come in. Once the characterization of the core features of the disease are in hand, one can make general claims and predictions more confidently. The other strategy involves comparative reasoning. Here, a researcher will compare two or more cases that share some salient feature but are also quite different in many other ways. For example, one could compare two instances of financial crises from different time periods and different parts of the world. By isolating common factors against a backdrop of differences, one can produce a plausibly generalizable explanation for at least some more general set of examples; for instance, one might identify a certain possible cause of some financial crises without claiming to have found the cause of all.

Finally, even with only a single case, one might attempt to draw some more general conclusions. For example, Morgan notes that Athenian democracy has often been used as a case study from which authors attempt to understand democracy more generally. However, this is likely the most tentative method here. The important point is that these case studies are often 'thick' in the sense that they include a significant amount of theoretical interpretation in the construction of the case narrative itself. That is, even internally, a single case study is usually informed and structured by theory. As such, this background theory is doing much of the work to support (tentative) generalization.

6.5.5 Summing Up

Any of the specific suggestions in this section might be worthwhile in some situations but not others. Nonetheless, it seems worth summarizing some general trends. First of all, there are legitimate methods that make use of extremely small sample sizes. Additionally, they tend to be most commonly applied when larger-sample statistical methods are unavailable or unfeasible. They seem most helpful when exploring questions about which researchers know comparatively little or have relatively low confidence. They can certainly play roles in discovery, but also can provide evidence for claims that are more general than the specific cases at hand. These generalizations are partial, highlighting some specific fact or feature of this case that might be important. Moreover, they work best with significant scaffolding: either significant theoretical background to support a plausible hypothesis about the significant factors in a given case, or rigorous mathematical models to test against, or through comparing multiple cases, or combining evidence from multiple sources. Confirmatory evidence from cases is likely weak, nonetheless. Arguably, this evidence is better for ruling out possible explanations, especially strong claims such as the claim that some factor is necessary for the phenomenon of interest.

6.6 Interpreting Small Samples in Comparative Psychology

As noted above, the standard foundation story for comparative psychology takes the field to have come into its own when it rejected the use of anecdotes in favor of experiments. However, with the frequent use of experiments involving extremely small samples, I suggest that we see these small-sampled studies as something in between the two, usually opposed, kinds of method. Small sample studies have some of the strengths that are usually ascribed to well-designed experiments, and some of the weaknesses ascribed to standard anecdotes. They provide stronger evidence than that provided by a one-off observation, but not as strong evidence as that provided by experiments with larger sample sizes. To illustrate more specifically, I return to the concerns lodged against anecdotes in Section 6.2. The question is whether, and how, these worries about anecdotes apply to small sample studies.

Concern 1: *Anecdotes can be cherry-picked to make a predetermined point.* In current practice, this worry might remain; researchers only publish the experiments that give positive results. As such, one might, for example, publish a paper showing an individual animal's impressive performance on a task without the context provided by previous failures on a swath of related tasks. This is a problem with publication practice in general (known as the file-drawer problem; Romero 2019), but it is arguably exacerbated by small samples, since flukes (of all sorts) are a higher risk. This is why I flagged the suggestion of registered reports above. If papers are accepted based on methods before experiments are done, then they won't be cherry-picked based on results. If this suggestion is implemented, it remains a worry that *existing* studies report cherry-picked experiments, though perhaps not to the degree of full anecdotes. This remains a reason to be cautious when interpreting the studies.

Concern 2: *We lack control over and knowledge of background conditions of anecdotes.*

Concern 3: *Anecdotes are narrative in structure, rather than providing analyzable data.* I'll take these worries together; small sample experiments pretty clearly avoid both. The conditions of a well-designed experiment are well controlled, even if the sample size is small. As such, one can have a relatively good sense of what the animal is responding to. Of course, the worries about underdetermination that have animated much of this book remain, but sample size itself doesn't affect this substantially. Similarly, small sample experiments do collect and analyze data. As such, worries about narrative structure, such as that it might prime anthropomorphic interpretations, are reduced.

A second reason to take these concerns together is that, at least for incident reports and case studies, narrative structure is precisely *how* researchers often include the kinds of details that identify possible causal factors responsible for the phenomenon. In this sense, the data reported in a small sample experiment might be *thin* in the sense that case studies are *thick*: they are able to convey likely causes without needing to, themselves, include substantial theory. This hurts the generalizability of small sample experiments. It has its benefits though, given my earlier concern that researchers should be able to use the experimental findings as a shared currency for debate, where disagreements are brought into the open in interpretation (Chapter 1; Dacey 2023). Nonetheless, the presence of data itself might create a misleading veneer of generalizability: it may *look like* the kind of data that we can straightforwardly generalize, but we should resist that impulse.

Concern 4: *Anecdotes are non-repeatable (non-replicable), and so can't be confirmed independently.* Here, things get more complicated. Small sample experiments have records of methods, which means that replication *attempts* are possible in a way that they would not be for anecdotes. It's less clear how many would replicate, though. The key here, for my purposes, is what features of the experiment one thinks ought to be resampled (changed) in the replication attempt (Machery 2021). In a within-subjects study, one can be confident that the effect would replicate if the same population of individuals were used (or the same individual, such as Ayumu). However, when the sample is small, there is little reason to think that the effect would replicate if individuals were sampled from the larger population of species members. I don't know what would happen if there were a large-scale replication project in comparative psychology, but I suspect replication rates would be disappointing if species populations were resampled. It is not as clear how to interpret failures to replicate as one might hope (Boyle 2021), so the issue would have to be settled case by case. Those worried about replication might still find it useful to flag the different evidential value provided by small sample studies which may have a higher risk of replication failure. Additionally, we should be careful to note that replication and generalization are separate (if related) issues.

Concern 5: *Anecdotes don't support generalization.* The question of whether, how, and to what degree small sample studies support generalization to larger populations is, of course, the central question of this chapter. I claim that they *can*, but the generalizations should be made much more like the analogues in the last section than the way experiments are usually generalized. That is, the result itself should not be directly generalized, though there might be lessons that can learned more generally. This will require interpretation of the finding and scaffolding with related findings and background theory, as with the methods above.

By contrast, Inoue and Matsuzawa summarize their findings by simply plugging generics in place of the specifics in a simple description of the study. So, "Ayumu outperformed the human average in this sample" becomes "young chimpanzees have an extraordinary working memory capability for numerical recollection better than that of human adults." Most strikingly, "Ayumu" (or, for that matter, any of the three young chimpanzees) is replaced with "young chimpanzees." This is not an uncommon way of reporting findings.[11] The authors probably don't intend to claim that this is true of all or even necessarily most young chimpanzees. But regardless of exactly what is meant by the appeal to "young chimpanzees" (not to mention what is understood by readers), it clearly extends the result beyond Ayumu without any consideration of why or how. This is a hasty, unjustified generalization because the sample size is too small.

What, then, do we learn from the candidate analogues above for a case like this? Firstly, more limited generalization may be reasonable. Most simply, "at least some young chimpanzees" might be fine. Secondly, rather than generalizing the claim about performance itself, there might be interesting theory-laden claims that could generalize—for example, claims about the cognitive systems involved. These sorts of claims, as in incident reports and case studies, must be scaffolded by other related findings or background theory. It takes interpretive and argumentative work to make that step. The examples above provide some guidelines for this interpretive work. In making the step to generalizable claims about cognitive systems, the first possibility comes from the idea that a case can *infirm* a hypothesis, even if it can't *confirm* any hypothesis. Here, the fact that one member of a species is able to perform a task to a certain criterion shows that it is *possible* for a member of that species to do so. Thus, any hypothesis that rules out the possibility of chimpanzees performing on that level is seriously challenged. Relatedly, we can mimic lesion dissociation logic: If Task A can be performed by an individual who cannot perform Task B, then it cannot be the case that the capacity responsible for performance

---

[11] This may be an interesting case for the literature on semantics of generic terms (e.g., Leslie 2008), though that is beside the point for now.

170 SAMPLE SIZE AND GENERALIZABILITY

of Task A is necessary for performance on Task B.[12] So perhaps we can conclude that successful performance shows that some capacity believed to be absent (say, language) is not *necessary* for performance on the memory task. However, there may still be lots of hypotheses that do allow this level of performance, and the finding cannot tell between them. This is progress, though perhaps limited: it's not clear what live hypotheses would have actually *ruled out* Ayumu's performance on the task.

Generalizability may also vary based on the expected variance within a species. If there is good reason to expect that members of a species don't vary much, generalizing might be safer. However, this should not be taken for granted, and an anthropocentric intuition that nonhuman animals all seem similar should not ground small sample studies. Similarly, some tasks will involve less variance and so, perhaps, more confident generalization. The difference in performance between Ayumu and the five other chimpanzees that began that study suggest (tentatively) that this particular case is one with high variance across individuals.

It's hard to say much more about the Inoue and Matsuzawa study here, because the experiment is not designed or purported to test between models of cognition. The goal here seems to simply be to see what the peak level of performance is.[13] Lots of experiments in comparative psychology are like this. As discussed in the last chapter, one can conceive of the project of discovering the limits of performance as simply a different project than the project of actually understanding how animal minds operate. I prefer to think the two are linked and am interested in how tests of limits can inform claims about cognitive mechanism. Perhaps, between two candidate models of chimpanzee working memory, one predicts better performance than the other. In such a case, perhaps the model predicting better performance would be supported by this result, but actually making that case would require a model that makes more specific predictions. Moreover, as suggested by vision science, more specific models may help scaffold small-N studies.

Even then, Ayumu might be doing the same thing as other chimpanzees but better, or he might be doing something entirely different. It would also likely require a larger set of related experiments with Ayumu to scaffold understanding of his performance in the first experiment. These could use other individuals as a way of fleshing out a general understanding of the capacities involved, but

---

[12] This basic inference structure is also employed in developmental psychology, though usually with larger sample sizes (Perner and Lang 1999).
[13] As such, this experiment is a prime example of the kind of 'trophy hunting' that Allen (2014) objected to in Chapter 3, which he argued is widespread; see also Chapter 5.

it is also an option just to test a single animal and flesh out an understanding of what they are doing. This would be more exactly like case studies, in which a lot of research is put into a single case. There is precedent here, in the lifelong study of animals with interesting or seemingly exceptional abilities, such as Koko the gorilla (Patterson 1978) or Alex the gray parrot (Pepperberg 2009).

While there are lots of suggestive hints for ways to generalize with small samples, there are relatively few concrete answers. This, I think, is OK. It is unlikely that there is a single answer, and we should make use of the tools at our disposal. In any case, recognizing extremely small sampled studies as such marks them off for more cautious generalization.

6.7 Conclusion

The simplest point here is to reduce the weight granted to small sample studies as evidence for general claims about the nature of nonhuman cognitive capacities. I have argued through the book that the field ought to reduce the evidential weight of individual experiments in general, to help move away from a pernicious 'critical experiment' framing that still too often pervades. As such, small sample studies may not lose as much ground against larger-sampled experiments as it might first appear. All of our sources of evidence are weak, so even weak sources are welcome.

The line between small sample studies and 'regular' experiments is likely not a sharp one, nor is it likely a constant one. The confidence one might have in generalizing might scale more or less linearly with sample size, and it presumably depends on the species and capacity being studied. As such, there isn't any particular number below which the label must be applied.[14] Nevertheless, it seems clear that lots of experiments in comparative psychology fit, and at least some experiments are clear cases.

There may be other general impacts on the field to recognizing a different class of experiment here. Because small sample studies, characterized as such, look more like exploratory research, this framing could benefit the field by encouraging more exploratory research and reporting of more varied behaviors. In effect, experimental comparative psychology might look a bit more like field ethology. For example, Stanton et al. (2017) presented raccoons with the Aesop's fable task, in which they can gain access to a treat floating on water by dropping stones in to raise the water level. They report that one of the raccoons

---

[14] Just as there is no one answer about how much to increase samples for replicability.

managed to get the treat, not by dropping stones, but by ripping the entire apparatus off the floor and dumping it out. A field that relies on registered reports, and recognizes the limitations of data from such small sample sizes, may include substantially more reports of behavior like this. There is value to that, as these behaviors, intended by the experimenter or not, do provide insight into the animals.

Most importantly, though, this framing encourages honest reporting of the significance of studies for understanding a species more generally. Extreme low sample size studies do have evidential value, but it is limited. They ought to be treated like experiments in some ways, and like anecdotes or case studies in some other ways. Making the category can help us use them appropriately.

7

Measuring Consciousness

7.1 Introduction

As you read this book, you are experiencing a number of things. You see the words, black against the white page. You feel the texture of the paper, or perhaps the cool smoothness of the tablet screen. You hear music or other ambient sounds. You may feel a draft or the sun on your back. You might feel pleasure (hopefully!) as you reflect on these experiences. Whatever the specifics, you occupy a specific subjective point of view, and there is *something it is like* to be you right now. The book itself, in contrast, has no such experiences. If you are reading on a computer, it doesn't have experiences either, as smart as it is. Unlike these other objects, you are *conscious*.[1] The nature of this capacity to have experiences, of any kind, in the first place, is arguably among the biggest mysteries remaining in the universe. As such, rich and deep literatures have developed on several questions about consciousness, especially in the last few decades.

In this chapter I discuss, specifically, the question of whether members of a given species of animals are conscious. Are they more like you or the book in this regard? This is known as the distribution question (Allen and Trestman 2020). To answer it, we need a way to *measure* consciousness. This is a methodological challenge which overlaps with the challenges discussed in earlier chapters but also, in many ways, presents its own unique difficulties.

The very idea of an experiencing subject is starkly different from the standard scientific picture of the world. Over the years, philosophers have attempted to capture this mismatch in several ways. Nagel (1974) argues that conscious experience ("what it's like to be a bat") is a subjective thing, which is beyond objective science. Levine (1983) argues that any reduction of a phenomenal experience to physical events leaves an 'explanatory gap' between the

---

[1] I use the term 'consciousness' to refer to this *bare capacity for experience*. This usage fits Block's 'phenomenal consciousness' (Block 1995). The term 'consciousness' is sometimes taken to connote some of the more elaborate features of human mental life. It seems impossible to avoid some unintended connotations with any choice of terms here, as there is something of a terminological quagmire. So, I simply urge readers to set those anthropocentric connotations aside, if they have them.

Seven Challenges for the Science of Animal Minds. Mike Dacey, Oxford University Press. © Mike Dacey 2025.
DOI: 10.1093/9780198928102.003.0008

174 MEASURING CONSCIOUSNESS

physical and the mental: even if we identify the neural processes underlying pain, that doesn't explain why pain feels the way it does. Chalmers (1995) calls the problem of how subjective experience arises in a physical world the 'hard problem of consciousness.'

For reasons like this, consciousness was long diligently ignored across scientific (and scientifically minded) fields.[2] Things have begun to change in the last few decades. Attempts at the scientific study of animal consciousness run into some practical implications of the big problems just outlined. In short, just about everything is up for grabs. Things are so uncertain in this area that we can't be confident that we know the right questions to ask, or have the right the methodology to answer them. We cannot see a conscious mind, so we need to know what we can look for.[3]

Still, there is work to do here. It is clear that at least some animals are conscious. Consensus has arguably grown that consciousness is widespread (perhaps universal) among mammals and birds, as well as octopuses (Godfrey-Smith 2016). Things are perhaps more contentious with reptiles (Learmonth 2020), fish (Key 2016 and commentary), and arthropods such as insects, crabs, and lobsters (Mikhalevich and Powell 2020; Birch 2022). There is a phenomenon here to study, and recognizing so provides new avenues for research and perhaps a richer science (Andrews 2020). In April 2024, a call for such research was codified by the New York Declaration on Animal Consciousness, signed by hundreds of experts in the field. Perhaps more pressingly, understanding animal consciousness has enormous practical significance for animal welfare and ethics. Learning that some species is conscious, in particular that it is able to experience pain and pleasure, is significant to its moral status on most moral theories.

As I have emphasized in other parts of the book, we don't have to answer the big questions to make progress. There are lots of more specific questions we can ask, at least to start. I take the distribution question to be a more tractable starting point than bigger questions about the nature of consciousness. I also take it to be one for which the general strategy developed through the book works well: slowly accumulating pieces of incomplete evidence that can raise or lower the plausibility of a given claim. To answer the distribution question, we need to know what kinds of observations or measurements can provide this kind of evidence. The goal of this chapter is to provide a framework

---

[2] As noted in Chapter 5, cognitive ethology is perhaps the exception here; e.g., Griffin (1976).
[3] It is arguably conceivable that any set of behaviors we can imagine could be produced by an unconscious being, even those indistinguishable from a normal human being; they could be a philosophical 'zombie' (Chalmers 1996).

for thinking about candidate measures and how they relate. It is unclear how we can find consciousness when we don't even know what it is. So, the initial orienting question of the chapter is: just how much theory do we need to develop measures of consciousness?

The case study for this chapter is consciousness in cephalopods, especially cuttlefish and octopuses. These present especially interesting examples because research has revealed them to be so sophisticated, but they are invertebrate mollusks that diverged from the line that led to vertebrates well before they possessed a brain. Their lives in the sea, their body plans, and even their neural architecture are hugely different from ours. Thought about consciousness has tended to take the human case as the paradigm and starting point. The alienness of cephalopods strains this approach to its limit. Nonetheless, the evidence that at least some cephalopods are conscious is messy but, I think, ultimately persuasive.

7.2 Cephalopods

Cephalopoda is a taxonomic class of marine mollusks including octopuses, cuttlefish, squid, and nautiluses. The discussion here will focus on octopuses and cuttlefish, which have been the best studied groups in the class. As a general matter, cephalopods have soft bodies. The only hard part of an octopus is their beak. They have eight arms which are lined with suckers, and, lacking skeletal structure, they can bend their arms at any point. As a result, their body might as well be fluid; they can adopt all kinds of shapes, including squeezing through a hole just barely bigger than their eye. Their skin has specialized mini-organs that allow for near-instantaneous, directly controlled, color change, sometimes in elaborate and dynamic patterns. Their skin is also lined with chemical sensors, akin to smell or taste. Most octopuses live on the sea floor, in dens or protected overhangs. Cuttlefish are less protean in shape than octopuses, as their mantle includes an evolutionarily internalized shell known as a cuttlebone (for a general sense, cuttlefish look a bit like thicker squid). The cuttlebone allows buoyancy, so they often live off the sea floor but usually stay close to it or in shallow water. Both octopuses and cuttlefish have eyes that evolved independently of ours. They share the general lens-and-detector architecture of vertebrate eyes, though they do not detect color, and instead detect the polarization of light.

Both groups also have large brains relative to their body size, though octopuses present an interestingly unusual neural organization. Roughly three-fifths of their neurons are in the legs. These distributed neural systems seem to have considerable computational ability, and the legs are capable of considerable independence. Octopuses are extremely active and curious animals, and they explore their environment extensively. This exploration, then, involves a kind of trade-off between autonomous probing by the arms themselves and occasional top-down control from the central brain.

Taking this together, cephalopods are so different from us in so many ways that they really strain any research project that proceeds from humans as a starting point. This is not unique to work on consciousness, but it is especially problematic here. We often use human performance as a kind of benchmark for various forms of intelligence, but we can still see that intelligence unfold in other behaving animals. Consciousness, we can only observe in our own personal case, and we can only get reports about it from other humans.

Moreover, arguments in favor of consciousness in other species have often pointed out the evolutionary relatedness of that species to humans. This approach is more effective the more closely related that species is. It might reasonably be considered decisive on its own with closely related species such as great apes (though there is plenty of other evidence for them as well). It is simply not available when it comes to cephalopods: any sophisticated psychological capacities they possess would have evolved entirely independently of our own.

Nevertheless, the idea that cephalopods are conscious has gained currency.[4] In the UK, cephalopods are considered 'honorary vertebrates' for purposes as research subjects and are afforded the moral protections that status provides. More recently, motivated by a report by Birch et al. (2021), the UK government placed cephalopods (along with decapod crustaceans, including crabs, lobsters, and shrimp) as candidate subjects of more general animal welfare law. The idea has also become significant in popular culture. In recent years, popular books and documentaries have touted the intelligence and charisma of octopuses especially.

In general, my sense is that the arguments which have had the most success at convincing people (researchers and laypeople alike) that octopuses are conscious have employed the rather blunt method of pointing out just how smart they are. Discussions of cephalopod consciousness usually point out how large their brains are, as well as how complicated their capacities for associative

---

[4] My goal here is not to argue for cephalopod consciousness. For that, see Mather 2008, 2021a, 2021b, 2022a, 2022b; Godfrey-Smith 2016; Ponte et al. 2022.

learning (Borrelli and Fiorito, 2008) and learning by observation of others (Fiorito and Scotto 1992), the fact that they play with objects (Mather and Anderson 1999), use tools (Finn, Tregenza, and Norman 2009), and remember which areas they have recently foraged (Jozet-Alves et al., 2013).

These arguments can work, to some extent, on their own. Consciousness itself seems like a form of cognitive sophistication, like intelligence. As such, animals that have one might be likely to have the other. There are more sophisticated versions of this argument as well. Mather (2008) argues that intelligent capacities such as these in octopuses are evidence of unified, integrated processing of information, which is commonly thought to be an important function of consciousness (more on this below). Even so, we should want more concrete and better-grounded measures. Fortunately, the distribution question has received increasing attention in recent years, including discussion of high-level questions such as the role of theory.

7.3 Theory-Neutral, Theory-Heavy, and the Hunt for the Middle Way

Birch (2022) describes approaches to the distribution question as falling on a spectrum between a "theory-heavy" approach and a "theory-neutral" approach. He argues against these extremes and proposes his own middle way—a "theory-light" approach—instead. The space of options that this sets up will be helpful in framing the discussion here.

Birch summarizes the theory-heavy approach as follows: "We start with humans, we develop a well-confirmed, complete theory of consciousness, and take this theory 'off the shelf' and apply it to settle the question of whether animals, in disputed cases, are conscious or not" (2022, 134). There are a number of theories of consciousness on offer (Van Gulick 2021; Seth and Bayne 2022). These theories each identify some mechanism or phenomenon as the key indicator of consciousness, which we could look for in animals. Birch rejects the theory-heavy approach on the grounds that the theories themselves are too uncertain (more on this criticism below). In response, Halina, Harrison, and Klein (2022) argue that his particular characterization of the theory-heavy approach would not likely gain adherents anyway. They argue that the goal in any attempt to study animal consciousness is to learn about consciousness more generally. As such, work on animal consciousness would feed back into the theory itself. These authors still endorse a "theory-heavy" approach, because they still start with a commitment to a theory against which evidence of consciousness is compared. Their version of the approach, though, is (arguably) more moderate than the one Birch initially described, and I'll adopt this more liberal usage of the term.

Birch presents the theory-neutral approach, in contrast, as an attempt to identify indicators of consciousness that require no theory at all. The straightforward version of a theory-neutral approach would simply take a behavior we believe to require consciousness in humans and look for it in a nonhuman animal. In short, Birch argues that such an approach will be smuggling in theoretical assumptions even if the researcher fails to recognize it: theory is inescapable.

For his part, Birch (2022) proposes the theory-light approach as a middle way. The idea is to use some theory, but only minimally. His proposed theoretical commitment is *the facilitation hypothesis*: "conscious perception of a stimulus facilitates, relative to unconscious perception, a cluster of cognitive abilities in relation to that stimulus" (140). For actual evidence, one would have to show *facilitation*, not merely the presence of some ability or behavior. This requires that one can show a difference between performance on a (purportedly) unconscious version of the task and a (purportedly) conscious one (see also Crump and Birch 2021). The theoretical commitments here are thin but are still doing important work.

Others have joined this search for a middle-way approach. Shevlin (2021), for example, calls his suggestion the *modest theoretical approach*, to indicate more theoretical commitment than Birch's theory-light approach. He characterizes this as a dynamic equilibrium between indicators of consciousness and theory, which mutually inform one another. Perhaps the most systematic attempt at a middle-way approach is Ginsburg and Jablonka's (2019) proposal of Unlimited Associative Learning (UAL). UAL subsumes a suite of specific capabilities which, they argue, can mark the point at which consciousness emerged in evolution. It is not intended to be a *theory* of consciousness, and they intend it to be compatible with any viable theory (UAL will receive considerable attention in Section 7.5).

This is the overall space I will work in. In the next three sections, I describe each main approach in turn: theory-heavy, theory-light, and then UAL specifically. Each approach does have its strengths and weaknesses. Once we see what these are, we can consider how best to retain strengths while limiting weaknesses. I argue, the best approach is not necessarily to forge a new middle way but to frame the way we think about relationships between measures and theory in a way that brings theory-heavy and theory-light approaches closer together.

7.4 Theory-Heavy Approaches

Theory-heavy approaches adopt a particular theory as a guide before approaching the distribution question. The benefits are relatively clear. A theory of consciousness will tell you what to look for, be it neural mechanisms, cognitive capacities, behaviors, or whatever. In this section, I will simply mention two candidates that are instructively different. I'll first describe each and where it tells us to look, and then discuss some concerns for each theory. This is merely meant as a taste of how things go, not a survey (see van Gulick 2021; Seth and Bayne 2022 for surveys of existing theories).

7.4.1 Global Workspace Theory

Global workspace theory takes consciousness to arise with information that is *broadcast* across processing units in the brain (Baars 1993, 2005, 2017). We can arguably identify neural mechanisms of broadcast in humans, which are typically taken to heavily depend on the cortex (Dehaene and Changeux 2011; Mashour et al. 2020). This theory motivates a search for similar networks and activity patterns in other species. Since the theory was at least initially developed at a cognitive level, it is possible that similar broadcasts could be identified in those species that do not share a cortex, especially those with functionally similar structures. However, this theory is generally taken to suggest that consciousness is present in a relatively limited range of species.

For example, Bernard J. Baars, one of the most prominent proponents of global workspace theory, and his coauthors strike a skeptical tone in their discussion of possible consciousness in octopuses: "Clearly, a strong case for even the necessary conditions of consciousness in the octopus has not been made. Nevertheless, the present evidence on cephalopod behavior and physiology is by no means sufficient to rule out the possibility of precursors to consciousness in this species" (Edelman, Baars, and Seth 2005).[5] On the other hand, Mather (2008) uses the global workspace theory to frame her argument in favor of octopus consciousness. She does this by extracting the general idea that consciousness involves a centralized integration of information and presenting organized behavioral complexity as evidence of integration. Mather's is a looser idea of the global workspace than Edelman, Baars, and Seth's (in fact, I'd argue,

---

[5] These authors do not identify global workspace itself as their standard, but the basic neural architecture they emphasize fits the architecture Baars suggests elsewhere.

180 MEASURING CONSCIOUSNESS

lots of theories of consciousness could have gotten her there), but it is a perfectly reasonable reading.

The difference between these readings of the view highlight a problem that comes when applying it to nonhuman animals. It's hard to know exactly which theoretical posits to generalize; is it a particular neural network pattern, or a more general set of functional properties of information processing? Moreover, this theory was originally constructed to explain conscious *access*, as measured by verbal reportability. Even if it is the right account of conscious verbal reports, there is less reason to think the broadcast network is necessary for more basic forms of perceptual or pleasure/pain consciousness (e.g., Lamme 2006; Block 2007). If it is not, global workspace theory would make a bad guide for the distribution question, because it only looks for one kind of consciousness.

7.4.2 Midbrain View and Reafference

Based on work involving humans with brain damage and deficit, Merker (2005, 2007) argues that the midbrain is sufficient for consciousness. This view would likely imply a considerably broader range of conscious species than the global workspace. While this has not been a popular account of human consciousness, it has had some uptake by researchers considering the possibility of invertebrate consciousness. These authors do not focus on the particular neural architecture of a midbrain but, rather, the capacities it supports.

Barron and Klein (2016) (and Klein and Barron 2016), for example, argue that insects are conscious on these grounds. More to the heart of our discussion, Godfrey-Smith (2019, 2020) does so for octopus consciousness.[6] These authors emphasize two capacities as crucial. Most important is the ability to track the difference between sensory changes that result from one's own movement and those that result from events in the world. This is known as *reafference* and arguably establishes a primitive self/world distinction (Jékely, Godfrey-Smith, and Keijzer 2021). Second is the ability to integrate that 'model' of the world with information about the current needs of the animal, setting up the ability

---

[6] I'm imposing my own terminology on Godfrey-Smith here. He prefers the term 'subjectivity' for bare experience and uses 'consciousness' to refer to more elaborate forms of human consciousness which are modifications of bare subjectivity. He may contend that my translation is not perfect, but it is good enough for present purposes. Terminological rough patches like this are inevitable, given the quagmire noted in note 1. Another quirk of Godfrey-Smith's view is that he thinks perceptual experience and pain experience can arise separately. So, he thinks octopuses have both, and insects have perceptual experience but may not have pain experience.

to evaluate and choose courses of action. These abilities are widespread in the animal kingdom: octopuses and cuttlefish pretty clearly possess them.

This view faces a different set of criticisms, as the process he describes may be too basic to be conscious (e.g., Hill 2016). In other words, the abilities Merker describes may be necessary but not sufficient for conscious experience. The same could be said of Godfrey-Smith's approach, which spreads consciousness more broadly than most in the field would agree to.

7.4.3 Theory-Heavy Lessons

Each of these theories, if adopted, would provide guidance on where to look for consciousness in animals. But each has problems. Their first-pass problems are arguably opposite: global workspace theory shoots too high, while Merker and Godfrey-Smith shoot too low. However, as potential guides for the distribution question, there are also big-picture problems that they share. Indeed, these are problems that arguably any theory-heavy approach will share.

Firstly, all theories of consciousness are highly contentious and uncertain. Because theory-heavy approaches presuppose highly uncertain theory, any claims they imply about animal consciousness must be highly uncertain as well. Secondly, theory-heavy approaches face a *specificity problem* in how to generalize the posit they identify as crucial (the architecture/ability/lifestyle/ whatever) from the human case to a given nonhuman species (Shevlin 2021; Birch 2022 cites a similar *demandingness problem*). Theories of consciousness have typically been formulated by studying humans first. So, even where the goal is to develop a cross-species theory, we need to understand how things might work differently in various other species. We would need to decide just how much difference, and what kinds of difference, actually matter:[7] How much, and in what respects, must the animal's brain architecture look like the human broadcast network? How much, and in what respects, must the animal's brain function like the human midbrain? In some cases, the key posit may be obviously present or absent. However, a theory developed for the human case cannot tell us what to do in *edge cases*, which do not unambiguously display the presence or absence of the crucial posit. Such edge cases are, I'd argue, likely on most (if not all) viable theories.

---

[7] This problem is closely related to the boundary problem of homology (Chapter 4) and the general problem of deciding what the requirements are of any cognitive process (Chapter 1). The worry here runs so deep that it (largely) leads Carruthers (2018) to argue that there is *no fact of the matter* about whether animals are conscious.

182 MEASURING CONSCIOUSNESS

With the key strengths and weaknesses of a theory-heavy approach in hand, and a more general sense of how a theory-heavy approach works, we can move to theory-light approaches for comparison.

7.5 Theory-Light Approaches

The idea behind theory-light approaches is that there might be indicators we could interpret as *measures* of consciousness which do not require substantial *theories* of consciousness. If so, we could identify conscious thoughts or behaviors and sort out the theory later. I'll discuss several candidate theory-light indicators, organized in two groups. The first includes putative indicators of perceptual consciousness, as opposed to unconscious processing of sensory information. The second includes putative indicators of pain experience, as opposed to unconscious damage detection (nociception). The list of indicators of perceptual consciousness is adapted from Birch (2022), while the list of markers of pain experience is adapted from Birch et al. 2021 (their list is itself adapted from Smith and Boyd 1991). Both lists are condensed, and each has an additional item added. Here again, the goal is more to give a feel for things than to systematically survey the options.

7.5.1 Perceptual Consciousness

7.5.1.1 Multisensory or Cross-Modal Learning

Palmer and Ramsey (2012) argue that the learning of cross-modal associations, for instance between a sound and a smell or between an image and a sound, requires conscious awareness of the stimuli. The necessity claim is controversial (Scott et al. 2018), but even failing that, it may still be that conscious processing at least facilitates cross-modal learning. Octopuses have shown some evidence of cross-modal learning. For example, Anderson and Mather (2010) tested octopuses' learning to open a jar to access food. They found that the octopuses got significantly faster at opening the jar only when presented both visual cues (seeing the enclosed crayfish) and chemical cues (herring mucus smeared on the outside of the jar). Neither cue alone seemed sufficient, though the performance improvement transferred to trials with visual cues only (see Maselli et al. 2020 and Mather 2021b, though, for some possible complications and discussion).

7.5.1.2 Rapid Reversal Learning

Travers, Frith, and Shea (2018) asked participants to indicate on which side of a computer screen a target was presented. They preceded the target with arrows pointing towards the side of the upcoming target. Unsurprisingly, people responded faster when given this prompt. However, after extensive training, the researchers made the arrows less reliable, or even reliably indicating the opposite direction. In one condition, the arrows were 'masked' by being covered by a new stimulus so quickly that participants didn't report having consciously seen them. In the masked condition participants did not correct for the change; however, when unmasked, the new contingency was learned quickly. This quick reversal of a learned association seems to require conscious perception of the stimuli.

Going back to the 1950s, experiments have demonstrated reversal of a visual discrimination association in octopuses (see Bublitz et al. 2017 for some criticisms of this literature). Mackintosh and Mackintosh (1963) even found that octopuses made the appropriate reversal more quickly as it was repeated. More recently, Bublitz et al. (2021) demonstrated repeated (serial) reversal in a spatial discrimination task. In this task, octopuses were rewarded after touching one of two feeding tubes on either side of a tank. The octopuses learned the task and were then able to react when the tube that gave a reward was reversed, and even did so more quickly with repeated reversals. In total, reversal learning in octopus seems established. However, a version of the experiment that uses a cue that can be 'masked' is necessary for the strongest comparison to the human case. It is admittedly hard to see how that kind of masking is possible with octopuses. But even in the absence of the direct comparison, the bare ability of reversal learning could be taken as some, weak, evidence of consciousness.

7.5.1.3 Neural Correlates of Consciousness

There has been a long hunt for neural activities that correlate with conscious perception in humans. In work using electroencephalography (EEG) to measure field potentials in the brain, two signals that have gotten attention are the N200 in the occipital and parietal lobes (Rutiku, Aru, and Bachman 2016), and the P300 in the frontal lobes (Del Cul, Baillet, and Dehaene 2007). Observed correlations, however, cannot differentiate between activities that actually underlie consciousness and activities that simply precede or follow those (Koch et al. 2016). It's also difficult to know how to compare signals across species, given differences in brain size and architecture, especially in a brain as different as a cephalopod's. Even so, Bullock and colleagues have

184 MEASURING CONSCIOUSNESS

argued that field potentials (measured with inserted electrodes) are broadly similar to vertebrates in both octopus (Bullock 1984; see also Gutnick et al. 2023) and cuttlefish (Bullock and Budelmann 1991). So, this approach could have promise if we can figure out how to generalize the important brain signals (though this is no small problem).

7.5.2 Pain Experience

7.5.2.1 Protection of Damaged Body Parts
This includes guarding, rubbing, or otherwise tending areas presenting damage or noxious stimulus. This demonstrates the ability to recognize a specific location of the noxious stimulus, as well as motivational roles for processing that information. There is considerable anecdotal evidence of this kind of wound tending in cephalopods. More systematically, Alupay, Hadjisolomou, and Crook (2014) found that octopuses tended to areas that had been damaged but not those that had been merely touched. They also found that damage to the tip of an arm that had been removed led to sensitization (on a shorter term than intact animals) but not any apparent tending behavior.

7.5.2.2 Motivational Trade-Offs
This is the ability to trade off putatively 'painful' stimuli against rewards in a flexible way. The goal is to demonstrate the integration of nociception information with other information (e.g., Sneddon et al. 2014; however, see Irvine 2020 for critical review). This has not been tested directly in cephalopods, but there may be promising approaches. Ross (1971) found that octopuses learn to avoid hermit crabs with stinging anemone on their shells (which formed a basis for several training procedures). McLean (1983) further observed that octopuses approached these crabs in cautious, flexible ways, such as approaching slowly with one arm, reaching below the anemone, and blowing jets of water at it. None, however, tested directly for trade-offs between an aversive stimulus such as the anemone sting and the value of the food source.

7.5.2.3 Valuing Putative Analgesics/Anesthetics When Injured
Crook (2021) performed an experiment demonstrating that octopuses prefer locations at which analgesics were delivered. This arguably presents the strongest current piece of evidence for octopus pain. Crook applied noxious stimuli by injecting acetic acid (versus a saline solution control), and injected lidocaine as an analgesic. The octopuses injected with acetic acid in their

preferred chamber would switch preference away from that chamber. In turn, though, they would switch to prefer a chamber in which lidocaine was injected, but only if previously injected with acetic acid. So, lidocaine only made a difference if there was a noxious stimulus to alleviate (Crook also reported tending of the injured arm in individuals injected with acetic acid but not lidocaine).

7.5.2.4 Mood-Like Behavior Patterns

Some sustained, global behavior patterns look a bit look basic forms of moods. These patterns last well after an aversive/positive stimulus and affect responses to all kinds of stimuli. Nociceptive sensitization in gastropod mollusks (a group including snails and slugs) looks, in this way, like fear (Godfrey-Smith 2020; Crook and Walters 2011). Positive and negative biases in behavior can be observed after receiving unexpected rewards or punishment. These are observed in bees (Bateson et al. 2011; Solvi et al. 2016), and Giancola-Detmering and Crook (2024) have recently shown a negativity bias in the interpretation of an ambiguous cue following stress in cuttlefish.

7.5.3 Indicators and Measures

All of these candidate measures of consciousness are uncertain. In the particular case of octopuses, it is not clear whether any are present in the way that would provide the strongest evidence for consciousness. More broadly, though, all of these behaviors and neural features can be explained in ways that do not attribute consciousness. In the language of measurement theory, potential indicators face a *coordination problem* of measurement (Tal 2017). If a candidate indicator actually tracks the thing that is intended to be measured, the two are *coordinated* (or, the indicator is *valid*). In this literature, an "indicator" becomes a "measurement" when it is interpreted as such. So the question of whether a measure is coordinated with its target requires thinking about the justification of that interpretation.

In a theory-heavy approach, that interpretation is grounded in the theory, so the question whether the indicator is coordinated is a question about the justification of the theory. Theory-light approaches, in contrast, mostly base their claims on the correlation between the indicator under question and another indicator: usually subjective reports of conscious experience in humans. So we would track performance on the behavioral task, or the neural activity, against the participants' reports of what they experienced or did not experience (on some reporting scale). This is true, even if indirectly, for the measures described above. For example, we know that pain motivates wound tending in humans because we can ask people why they are tending some part of their body ("it hurts!"). Similarly, humans in the unmasked reversal learning experiment were able to report their awareness of the cue, but not in the masked condition.

The question, then, is whether subjective reports are *themselves* coordinated with conscious experience.[8] While it might seem natural to assume that they are, methods based on introspection have long faced criticism, for good reason (Irvine 2012, 2021b). For instance, any subjective report measure might be influenced as much by participants' interpretation of the question as by their actual experience (Michel 2019). This is exacerbated by the fact that the measures would need to remain coordinated *across species* (Browning and Veit 2020). As with any of these examples in the book, animals may perform tasks in ways that are fundamentally different from humans.

Theory-light approaches also fail to tell us what to make of unexpected patterns of results: suppose members of some species demonstrate "conscious" behaviors according to one measure, but not others. As he articulates his theory-light approach, Birch (2022) asserts that the measures he names will cluster, even though he doesn't expect perfect correlation. This is, I take it, an empirical bet based on the unstated assumption that there is a single phenomenon, consciousness, that they collectively measure. However, Halina, Harrison, and Klein (2022) argue that, without some more substantial theoretical grounding, these measures struggle to give guidance on *edge cases* that display some of the indicators but not others: how many need to be present to declare consciousness? Are some indicators more important than others; if so, when and why?[9]

Overall, both theory-heavy and theory-light approaches have problems with justification and how they deal with edge cases. Perhaps, then, the right move is further to the middle, seeking some of the justificatory scaffolding of theory, without overcommitment to uncertain theoretical detail. Unlimited Associative Learning (UAL) might be thought to provide just such an approach, so I turn to it now. In short, I don't think it does provide such a middle way, but the way that it fails is instructive for my positive proposal.

---

[8] In the sense that measurement of consciousness is in a similar state to early work on the measurement of temperature, a landmark example which may be a fruitful comparison (Chang 2004; Michel 2019).

[9] More on this objection below (Section 7.7).

7.6 Unlimited Associative Learning

Ginsburg and Jablonka describe UAL as an 'evolutionary transition marker.' This means that they think UAL requires consciousness and is likely among the earliest-emerging abilities that do. So, mapping the distribution of UAL across species likely provides a good approximation of the distribution of consciousness. It is not a theory of consciousness itself, and they actually intend it to be compatible with most theories in the literature (whether this is actually true is contestable).

As they describe UAL, "A system with capacity for UAL can, within its own lifetime, learn about the world and about itself in an open-ended way . . . the possibilities are sufficiently open-ended that there is no serious prospect of all of the possible associative links it can produce being formed by a real organism with a realistic, finite lifespan" (Birch, Ginsburg, and Jablonka 2020, 8). They contrast this with "Limited Associative Learning" (LAL), which is an earlier-emerging suite of abilities that they believe does not require consciousness. As they describe LAL, "We define LAL as conditioning that includes both self-and world-learning [near enough, operant and classical conditioning] and involves the formation of predictive relations between noncompound stimuli, actions, and reinforcers. 'Noncompound' refers to features of objects or actions that cannot be fused into a single nonadditive discernable pattern" (Ginsburg and Jablonka 2019, 321). Ginsburg and Jablonka argue that consciousness emerged in the evolutionary transition between the suite of capacities present in LAL and those present in UAL.

To decide what specific capacities UAL includes, they start with a set of "hallmarks" that they take to be individually necessary and jointly sufficient for consciousness, according to all viable theories. The hallmarks include global accessibility or broadcast across processing units, binding of perceived features into objects and locations, intentionality, integration over time, and registration of a self/other distinction, among others. These hallmarks, in turn, suggest candidate indicators. For instance, the learning abilities noted above—multimodal learning and rapid reversal learning—would be examples of learning that UAL makes possible but LAL does not (for a full list see Birch, Ginsburg, and Jablonka 2020, table 1, pp. 6–7).

Because they take the hallmarks of consciousness to hold for any viable theory, Ginsburg and Jablonka do not take themselves to be committed to a specific theory. However, they do describe a basic mechanistic architecture that they take to be necessary for UAL (e.g., 2019, 361–2; Birch, Ginsburg, and Jablonka 2020, 49). This involves a complicated set of interactions between memory systems, systems that integrate sensory stimulation and motor information, and what they call a "central association unit." They note that these systems or "units" need not occupy specific physical locations; they are characterized functionally. Thus, presumably, the general architecture could be implemented in many different ways (perhaps as described by a given theory such as the global workspace or midbrain theory).

The (purported) ability to accommodate any viable theory of consciousness might mean that UAL can avoid challenges that arise for theory-heavy approaches because of the uncertainty of specific theories of consciousness. The sketch of a mechanism might also justify the choices of indicators and explain why that that set of capacities are expected to come together and potentially help with edge cases (e.g., Halina, Harrison, and Klein 2022). Unfortunately, I think the idea that UAL can play this middle-way role is based on a misreading of the explanatory role of associative models (see Chapter 3). To set this point up, I'll start with a worry raised by Irvine (2021a) in response to UAL.

Irvine argues that each kind of associative learning that Ginsburg and Jablonka describe as a feature of UAL is merely an operationalization of one of the hallmarks they list. For example, multisensory associative learning is just an operationalization of global broadcast, and associative learning of compound stimuli is just an operationalization of feature binding. Irvine frames this as a criticism, though she doesn't think it poses much of a practical problem. My reading is that Irvine has accurately diagnosed exactly what UAL does and, thus, its strengths and weaknesses. Perhaps this is a different role than Ginsburg and Jablonka intend, but it may still have value.

Here's why I say this: associative models, as I have argued, are abstract, partial descriptions of the processing involved (Chapter 3; see also Dacey 2016b, 2017b, 2019b). Associations are not, as they are often thought to be, a particular kind of processing; they are features of lots of kinds of processing.

To say that two representations are associated is just to say that they tend to come together. Most of the hallmarks of consciousness that Ginsburg and Jablonka start with are different ways that information needs to be connected or integrated. As a general matter, many of the interesting features of consciousness have to do with the bringing together of information from different places—from difference senses, from different locations in the perceptual field, from different processing units in the brain. To say that two pieces of information are associated is precisely to say that they are integrated, in some way at least. Associative models are very flexible descriptive tools which can describe many kinds of connections between representations in many settings. It should not be a surprise that association, so understood, could be a valuable way to think about these phenomena of consciousness. However, we need to be clear that identifying an association is simply stating the *fact* of integration; it is not explaining how they are integrated or identifying the mechanism that accomplishes it.

Associative learning paradigms are relatively well-known and provide relatively direct means of testing for integration. Thus, the value of characterizing these phenomena in the somewhat unified terms of association is that it brings them closer to experimental testing. This is, to my mind, tremendously helpful, and it is exactly the kind of work associative models do well. So, when it comes to the particular candidate measures, I think working in associative terms is helpful. However, this significantly shifts how we should see UAL as a construct and the work it can do.

If we reject the idea that associative processing is a single, specific *kind* of processing, redescribing things in associative terms guarantees very little for the underlying mechanisms. Thus, the idea that UAL describes a particular unified capacity is undermined. It collapses into a theory-light approach where each of the 'hallmarks' of consciousness independently plays a justificatory role for some measures, akin to Birch's facilitation hypothesis. This is basically the approach I will suggest, but to get a better sense of what that means and what it doesn't, I'll consider another objection to UAL.

7.7 Clusters and Clutters

Halina, Harrison, and Klein (2022) criticize UAL, in its more theory-light form,[10] on the grounds that the various features it names as indicators of consciousness form a mere "clutter" rather than a "cluster." A "clutter" is a disjunctive collection of measures, whereas a "cluster" would have some unifying principle to tie them together. I see three main complaints that Halina, Harrison, and Klein raise against a theory such as UAL that merely identifies a "clutter" of features or indicators. The first was already mentioned (Section 7.5): they struggle to handle edge cases, in which some but not all indicators are present (or perhaps some are incompletely present). They also assert that such a "clutter" of indicators is epistemically "unsatisfying" and cannot explain why *this* set of indicators is on the list and not some other set. I take this to be separable into a worry about explanation and a worry about justification.

---

[10] They note that it has slid between more theory-heavy and theory-light formulations in different publications.

190   MEASURING CONSCIOUSNESS

Thus, their second complaint is that UAL cannot *explain* the relation that any given indicator has to a larger construct of consciousness. The third is that it cannot *justify* the inclusion of any given indicator on the list or the exclusion of some other candidate.

Halina, Harrison, and Klein take the "cluster" versus "clutter" distinction from a discussion of the concept of innateness by Mameli (2008). In that discussion, Mameli is addressing the definition of innateness: that is, the "clutter" would be a collection of defining features of innateness. Based on work from the philosophy of science on 'natural kinds' (e.g., Boyd 1991), various theories have been proposed which take kinds to be defined by a *homeostatic property cluster*. This is one approach we could take towards innateness, or UAL, or even consciousness itself: a cluster view of consciousness has been proposed by Bayne and Shea (2020). The criticisms levelled by Halina, Harrison, and Klein are compelling if the goal of UAL is to define such a construct. If we are discussing the defining features of a construct, we want a cluster rather than a clutter.

This implies that the mistake of UAL is to gather together the various kinds of associative learning and take them to *constitute* a particular kind of system that is likely to also be conscious. As I've argued, it can't do this work. It may be that cluster views are often right, even for consciousness itself. However, we should not make the mistake of identifying the features that we use as convenient indicators of consciousness with the features that actually constitute conscious systems.

For a fuller sense of what I mean, compare the search for consciousness to the search for planets orbiting other stars (exoplanets). Astronomers searching for exoplanets are in the position of looking at stars which appear, from a great distance, as simply points of light. Though the details are, of course, very different, they are in a similar position to psychologists looking for animal consciousness in that both face substantial epistemic limitations because of their position relative to the target phenomenon. There are two main ways that exoplanets are discovered (Wright and Gaudi 2013). First, if the planet's orbit takes it in front of the star (from our position on earth), we can observe periodic dips in the intensity of light coming from the star. Second, we can observe small wobbles in the position of the star caused by the orbiting planet (properly, the star and planet both orbit their shared center of mass) by detecting slight periodic Doppler shifts in the light coming from the star. These are reliable methods (or so we think), but no one would mistake them for *constituting* the construct of a solar system. I urge researchers (especially theory-light proponents) to look at candidate measures of consciousness similarly. If we do, then Halina, Harrison, and Klein's criticisms have much less bite: what matters is that the set of measures is coordinated, useful, and open to revision when necessary.

There are two different notions of "cluster" that could be at play if we think about measures this way, and I propose we separate them. The first is when things actually do cluster in the world as an empirical matter; that is, the appearance of one indicator will (somewhat) reliably correlate with the appearance of others. I'll call this an *empirical cluster*. The second is when indicators form a kind of cluster because they are all based on shared assumptions about consciousness or shared methodology. I'll call this an *epistemic cluster*.

I take one key difference between theory-heavy and theory-light approaches to be in how they engage these different kinds of cluster. A theory-light approach is likely to start with one or more epistemic clusters and hope that research eventually bears out that they are also empirical clusters. A theory-heavy approach is likely to assume that any viable set of epistemic clusters must also be empirical clusters, because of larger theoretical commitments.

The approach I advocate here is not, exactly, to propose a new middle way between theory-heavy and theory-light approaches, as UAL (on my reading) attempted to do. Instead, the aim is to bring theory-heavy and theory-light approaches closer together by thinking more carefully about how to engage with these different kinds of clusters. I call this the *multiple clusters* framework.

7.8 Multiple Clusters

The overall idea is that researchers addressing the distribution question should start by assembling epistemic clusters of measures and working to establish which measures are actually best. This involves seeing which set of measures forms the right kind of empirical cluster and developing theoretical understanding of the measures and of consciousness. A healthy version of this, I suggest, will include multiple clusters based on different theoretical/methodological assumptions, each cluster including multiple (candidate) measures. This may more obviously fit a theory-light approach than a theory-heavy approach, but I argue it fits both. More importantly, I argue that it can help reduce the worries with each approach noted above—lacking justification and difficulties with edge cases.

I address theory-light approaches first (for an example of a theory-light approach as advocated here, see Figure 7.1). Theory-light approaches often identify specific candidate indicators by noting tasks for which conscious processing seems to make a difference in humans. They are then interpreted with some theoretical assumptions (in Birch's (2021) terms, this is why they are theory-light instead of theory-neutral). A measure can fail at both points; it may be that it is not actually a coordinated measure, and the theoretical assumptions supporting it may be wrong. As such, the candidate indicator might be explained by an unconscious process rather than a conscious one. Because of this, it is hard to know which measures are actually justified. The idea of having multiple clusters is to scaffold multiple, (reasonably) independent, lines of evidence into a collective argument for or against consciousness (e.g., Browning and Veit 2020). If we find evidence of consciousness in a species that is based on several independent theoretical/methodological assumptions, then the case is less vulnerable to the fact that any one such assumption might be mistaken. We can, in effect, triangulate the phenomenon from multiple different directions. Having multiple (candidate) measures within each cluster makes the cluster itself more robust: an individual indicator might return idiosyncratic results in a given case, or may even be impossible to measure in a given species.

This framework addresses edge cases by recognizing that the clusters and the indicators in each cluster are preliminary and subject to revision. The indicators that we include in a given cluster can shift, as new indicators are added and old ones dropped. We might also realize that a purported cluster is not on solid ground and eliminate it, or develop new clusters to include. In the

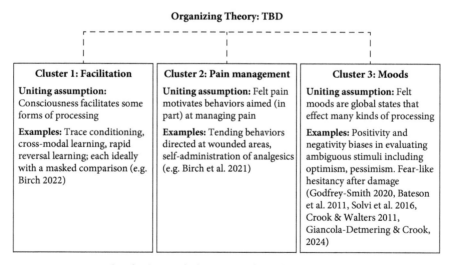

Fig. 7.1 An example of a theory-light approach applying the multiple clusters framework. "TBD" means "to be determined."

meantime, if the measures in a given cluster are giving inconsistent results, we can consider the evidence to be partial; these are epistemic clusters rather than defining features of a kind of system.

The same basic process works for theory-heavy approaches (for an example of a theory-heavy approach as advocated here, see Figure 7.2). An important step for a theory-heavy approach would be to identify different epistemic clusters that follow from the theory. Theories of consciousness tend to include a number of commitments that may suggest different places to look for evidence of consciousness, either from different parts of the theory which are in-principle distinguishable or from different ways of making the theory precise. These assumptions can then be separated out and can each ground an epistemic cluster.

The justification problem for any theory-heavy approach is amplified to the extent that they put all their eggs in one basket. If one leans to heavily on a specific version of a specific theory and it turns out to be false, then the approach utterly collapses. This is a problem when all of the theories are as uncertain and contentious as they are. In my framing, though, we have some more grounds to assess different parts of a theory independently. For example, we might object to particular claims about the neural implementation of consciousness that are derived from some theory (e.g., cluster 1 in Figure 7.2) without objecting to overall functional claims (cluster 2 in Figure 7.2). Thus, the approach can be more robust to specific criticisms of a theory. The response to edge cases is

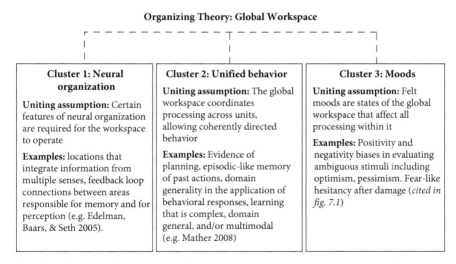

Fig. 7.2 An example of a theory-heavy approach applying the multiple clusters framework.

194   MEASURING CONSCIOUSNESS

much the same. The problem here for theory-heavy approaches was that they can vary in just how specific they require any claims to be. For instance, with global workspace theory, we can ask just how similar the neural structures involved need to be to the human case. In the multiple clusters framework, we might have multiple specific ways of making a theory more precise which we may take as different clusters (that is, we may have multiple versions of, or alternatives to, cluster 1 in Figure 7.2).

Aside from limiting these concerns, this framework does three more things that I take to be positive. First, it forces the articulation of assumptions behind the epistemic cluster. Second, it scaffolds responses as empirical results come it. Third, it brings theory-heavy and theory-light approaches closer together in ways that I hope can be fruitful. I'll say something about each of these in turn.

First, this approach requires that that the specific assumptions underlying candidate measures be explicitly articulated, because they form the basis for each cluster. These assumptions are not always as clearly and precisely articulated as they could be, by either theory-heavy or theory-light proponents. Theory-light proponents are probably more likely to leave assumptions unarticulated, while theory-heavy proponents are more likely to fail to be precise about which parts of the theory motivate a given measure. Articulating the epistemic clusters will make the assumptions more explicit, make the justification for a given set of indicators more concrete, and make debate about them more transparent.

Second, this framework helps scaffold theoretical development as new results come in. For a theory-light approach, this would be the project of gradually building up to a theory of consciousness. As research progresses, the assumptions behind any given cluster of indicators might be built up into something more substantial. We might trace theoretical connections between different epistemic clusters as we come to understand them better. For a theory-heavy approach, the process would instead be one of theory revision, as clusters and indicators shift and our understanding of them changes.

Third, this approach brings the theory-heavy and theory-light approaches closer together. The overall project for each approach has the same shape; researchers in each group are just coming at it from different directions. This shared shape also gives more concrete focuses for disagreement. Specifically, agreement and disagreement can be focused on a specific candidate indicator, cluster, or clustering assumption. The fact that the clusters themselves are separated from the theory (or lack of theory) means that they can be discussed without necessarily taking sides on the overall theory; too often discussions of the distribution problem immediately collapse into arguments about the theories rather than discussions of the animals themselves. This also gives more room for the possibility that theory-heavy and theory-light approaches can identify the same clusters or indicators (note that cluster 3 is substantially the same in Figures 7.1 and 7.2). Different researchers may frame them differently or place different weight, but that is also true of competing theory-heavy or theory-light approaches; it's not specific to the heavy/light divide. This proposal is no panacea, but at least, I think, the issues can be more specific and explicit.

7.9 Conclusion

Answering the distribution question, whatever one's preferred approach, requires finding measures that can tell us whether consciousness is present in any species under question. We need ways to measure consciousness. The framework of multiple clusters, I argue, can scaffold the project of identifying measures in fruitful ways. It doesn't tell us how much theory we need, or how to handle edge cases, or which measures are actually justified. But it brings these questions to the fore, and it structures discussion of them as things progress.

Back to cephalopods, the evidence for consciousness is not absolute or univocal. Many of the behaviors discussed above have been only partially observed, or observed in conditions with debatable interpretation. It is a messy business to gather data and interpret it from an animal so unlike us as a cephalopod. Moreover, it is not entirely clear whether these behaviors require consciousness in cephalopods: even if consciousness is involved in these behaviors for humans, cephalopods might do them differently. One's own theoretical stance will also change the way one interprets the evidence.

Nonetheless, I take the evidence for consciousness in at least octopuses and cuttlefish to be pretty convincing. I have not systematically surveyed the evidence here, but there is evidence from multiple different epistemic clusters of measures. Each piece of evidence is small. But taking them collectively, I think the more likely explanation is that these animals are conscious than that they are not. This is a defeasible claim, if these results are shown to be unrepresentative, or if it turns out they are generally based on bad theoretical assumptions.

The stakes in such a discussion might seem high. First, the moral significance of conscious experience is clear. This motivates the study of animal consciousness and arguably should motivate a willingness to make claims about it even when things are not certain (Birch 2017). In the other direction, it may seem daunting to make such a big claim as the one that some species entirely unlike us is conscious, given that the issues are so hard to think about and when one's own knowledge is so limited. To close, I step back and attempt to lower these epistemic stakes.

Through much of the book I have articulated ways that researchers can make and defend tentative claims about animal minds. In short, we might make a claim but express low certainty in it; perhaps the claim is more probable than its competition, but far from certain. Nor need these claims imply any kind of rich *understanding*. The likely weirdness of animal consciousness means that we may never really understand or imagine it. But that doesn't mean there is nothing we can say.

It's also worth remembering that, for all the abstract theory, the phenomenon here is concrete, embodied in individual living animals. For example, Godfrey-Smith (2019, 2020) focuses on *subjects*, or conscious beings, rather than consciousness-as-an-abstract-thing to bring the conversation back down to earth some. Along similar lines, Wemelsfelder writes, "To enter this conceptual arena is thus to ask, not '*what is consciousness*', but '*who are you*'?" (Birch et al. 2022, 22). These authors may or may not find my approach satisfactory, but their insight helps us think about what it means to make a claim about animal consciousness.

Even so, consciousness remains the deepest challenge for the science of animal minds. In a sense, all of the other challenges from the last six chapters apply to research on consciousness, only further shrouded by the veil of subjectivity. Researchers are making progress. The multiple clusters framework can help. But regardless, it is likely to be slow, difficult, and piecemeal, as is much of the work of understanding animal minds. In the short term, here as elsewhere we should be modest in our claims. We may never truly know what it is like to be a bat, but we can answer smaller questions about consciousness. Any progress we make is progress in better getting to know who is out there, sharing our world with us.

Conclusion: Of a Different Mind

8.1 Returning to the Big Picture

The seven challenges remain significant for the science of animal minds and are unlikely to be entirely resolved any time soon. These are intrinsic to the subject of study, each for different reasons. Nevertheless, there are ways of viewing and addressing them that are more likely to be productive than others. We cannot simply dismiss the challenges as inevitable, or simply dismiss any research that fails to completely avoid them. We need to face up to these challenges. In doing so here, I hope to have helped make some progress on each of them individually.

There is also a big-picture view that has run through the discussions of specific challenges. The approaches I have suggested collectively provide a rough sketch of a different structure for the science. In this Conclusion, I step back to present a more coherent picture of this positive view. I compare it to current 'standard practice' and collect specific suggestions for each stage of the scientific process.

The central idea of the approach is that substantive models of animal minds can only be evaluated in a holistic inference to the best explanation that makes use of evidence from everywhere we can get it, including related experiments, observations, and background theory. The watchword for this framework is *modesty*. I mean that in the sense that we should be modest about the amount of work that any single *piece* of the science can do. 1) No single *experiment* can decide in favor of, or can completely rule out, any substantive model. We need to accumulate and organize evidence across experiments to evaluate models. Some experiments, with especially small samples, should be interpreted more like case studies that provide indirect, theoretically scaffolded, evidence. No single measure for consciousness should be seen as individually decisive. 2) Individual appeals to *parsimony* can only provide partial evidence. They cannot set defaults that create unreasonable bars for experimental evidence to clear. 3) *Models* should be allowed to be partial descriptions of a process that may not describe everything the process does, and may not predict its behavior

Seven Challenges for the Science of Animal Minds. Mike Dacey, Oxford University Press. © Mike Dacey 2025.
DOI: 10.1093/9780198928102.003.0009

198 CONCLUSION

in every task and context in which it is engaged. 4) *Claims* about animal minds in general should be made in ways that reflect uncertainty; this includes big claims about cognition, consciousness, homology relations, or any generalization from the lab to the world, or from a lab sample to a taxonomic group.

Taking these together, individual research programs will typically operate with the aim of filling in one part of a large, multifaceted picture of animal minds, even with respect to a single capacity in a single species. There is no single 'true' study of the mind, and many approaches are needed. Even so, tracing the consequences of this kind of modesty into specific parts of the scientific project suggests some general approaches at each stage that will help implement the approach. Specifically, this approach stands in contrast to a common 'standard practice' (to use Buckner's [2011] term) in comparative psychology, which has exacerbated the challenges here.

The form of standard practice I have in mind is built on the conjunction of four main methodological features. I presented these first in the Introduction, and each of them has come up in different ways through the book. I repeat them here:

a. **Morgan's Canon:** the general preference for models that describe simpler psychological processes.
b. **Associative/cognitive divide:** Treating the choice between competing hypotheses as a choice between one option which is associative and one cognitive, with a large gap in perceived sophistication between them.
c. **Success testing:** The framing of most experimental designs around the question of whether an animal 'succeeds' at some task.
d. **Null hypothesis significance testing (NHST):** A statistical method that frames experimental statistics around the rejection of a 'null' hypothesis.

Not all of comparative psychology works this way, but much of it does. These four features fit together very well; they provide an overarching structure to the science at every stage from experimental design to statistical inference to model choice. Each of these features has been criticized in the philosophical literature, in discussions raised throughout the book. These criticisms have animated my own arguments, and my suggestions build on this existing work.

These criticisms have also influenced many philosophers and philosophically engaged comparative psychologists. Even so, standard practice has been hard to dislodge. It has been resilient in part, I take it, because existing criticisms have tended to target one of the four features at a time. This piecemeal approach is limited in two interconnected ways. First, we simply lack an alternative framework that approaches the generality and systematicity of standard practice. Standard practice applies straightforwardly across subject matter, including tests of different species and different cognitive abilities. This overarching structure helps researchers (and granting agencies) understand the place of individual experiments in the larger context of the field. It functions in the sense that it keeps things happening. It also makes sense, both in that it is internally coherent, and that it has some arguments in its favor (or, more minimally, there are good historical reasons things work this way). For these reasons alone, it is hard to give up without having an attractive replacement.

Second, the four features align so neatly that they reinforce one another. This makes it hard to simply remove one without addressing the others. Indeed, I argue that they align *so* neatly that simply removing one feature won't do much: the remaining three features can, in practical effect, recreate the removed one. This claim requires explanation. I'll build up to it.

In Chapter 1, I discussed the ways that success testing combined with NHST reinforces the preference for simpler processes in Morgan's Canon. In success testing experiments, the statistical null is generally 'no effect.' 'No effect' is a 'failure' on the animal's part, because 'some effect' would be successful completion of the task (hence, 'success testing'). Failures are generally predicted by the hypothesis positing the simpler system, because that system is not 'smart' enough to accomplish the task. NHST has the effect of granting epistemic privilege to the null. Putting this together, success testing plus NHST creates a practice in which the hypothesis positing the simpler process is granted epistemic privilege. This is exactly what Morgan's Canon dictates. Taken collectively, then, these three features not only align but mutually support and reinforce one another. Morgan's Canon can give (seemingly) independent reason to think that success testing with NHST is on the right track, and success testing with NHST can give (seemingly) independent reason to think Morgan's Canon is on the right track.

The compound of success testing plus NHST plus Morgan's Canon has a similarly mutually supportive relationship with the associative/cognitive divide. Morgan's Canon counsels the acceptance of the available model which posits the simplest process, which is often an associative model. It also counsels the *interpretation* of those models that posits the simplest processes. This is part of the reason that associative models are taken to describe only very simple processes (a version of the "associative processing view" I rejected in Chapter 3). So, collectively, the features of standard practice place downward pressure on the perceived sophistication of processes described by associative models. This, then, reinforces the idea that there is a stark divide between complex cognitive processes and 'mere' associative processes. In the other direction, associative models are extremely flexible in the kinds of processes they can describe (recall from Chapter 3 that this is because of the formalism of the models, not the processes that they describe), which sets up dynamics that support Morgan's Canon and success testing. Associative models are broadly successful at making behavioral predictions in lots of species in lots of contexts. This is taken to be a justification for Morgan's Canon (e.g., Meketa 2014): simpler associative processes are taken to be more likely in any specific case because they are 'taxonomically ubiquitous.' If they are ubiquitous across taxonomy, they are likely to be present in *this particular animal*, which is not usually the case for a competing cognitive model. In addition, this drives success testing, because researchers are constantly trying to rule out a simple associative explanation of some behavior.

This is what I mean when I say the four features fit together. They align in the sense that they work in the same directions. This means they are mutually reinforcing in the sense that they usually support the same conclusions about animal minds in a given research context (and arguably magnify the impact that any one feature would have alone). It also means that they are mutually supporting in the sense that each provides seemingly independent justification for the others, if one is placed under criticism. I suggest, in fact, that this mutual support is so direct that if one attempts to remove one of them, the rest will recreate it, in effect if not in name. To justify this stronger claim,[1] I'll consider how dropping each of the features individually could play out in practice.

For instance, suppose we stop deliberately applying Morgan's Canon while retaining the other features. The way that success testing assigns the statistical null will recreate the preference for hypotheses positing simpler processes: that hypothesis will consistently get the epistemic privilege granted to the 'null.' It may be weaker than explicit application of Morgan's Canon. However, the large gap of the associative/cognitive divide means that preferring the simpler process has significant consequences for the perceived intelligence of animals under study. So, the newly recreated de facto Morgan's Canon might look in practice quite a bit like the old one (at least when it comes to model choice).

---

[1] One need not go all the way to this conclusion to accept my main point. My main point is that properly addressing standard practice requires a more systematic approach than just attacking one feature at a time. I believe that the claim that the features are mutually reinforcing and supporting is sufficient. Nonetheless, I think the 'mutually recreating' claim is true and important, so I proceed.

If we, instead, try to dissolve only the associative/cognitive divide, it would be effectively recreated as well. Success testing looks for complex processes that are able to 'succeed' at some task. This creates a situation where the competing models are usually ordered by posited complexity: the complex process can accomplish the task, while the simpler process cannot. Once the models are ordered this way, the existing features apply pressure that increases the gap in sophistication between them. Morgan's Canon applies downward pressure on the perceived sophistication of the simpler process (especially). The fact that there is a simple process which is a 'permanent contender' to explain behaviors in any given context then motivates 'trophy hunting,' where researchers attempt to display behaviors that no simple process could create. Trophy hunting searches for the most sophisticated behaviors possible, which applies upward pressure on the sophistication of cognitive hypotheses under discussion. Thus, different parts of standard practice work in opposite directions to open up a stark divide between simple and complex candidate processes.

We could also attempt to drop only NHST, for example by shifting to Bayesian statistics. This might not lead to the recreation of NHST quite as instantaneously or automatically as the other examples. But with certain setups, Bayesian statistics can work a lot like NHST in the ways that matter here. They can still establish nulls, or grant privilege to one hypothesis (e.g., Kruschke 2011). For example, implementing Morgan's Canon in a Bayesian test might mean setting high priors for the simpler process. If we do this in a success testing framework, the practical difference in which statistical method we use may not be big; either way, the candidate models are aligned by sophistication, and epistemic privilege is granted to the simpler process. This may be enough to set up the basic dynamics described so far.

Removing success testing is likely the single strategy with the least automatic tendency to recreate the structure. In such a situation, candidate models might not be ordered by complexity of posited process. However, if we need to use Morgan's Canon to adjudicate model choice when the experiments are not decisive (as I argued in Chapter 1, this is often the case), then we fall back into the habit of treating one model as describing a simpler process. Once we are there, the alignment between NHST nulls and Morgan's Canon that comes with success testing may tempt researchers to fall back into that pattern, so we would need to remain vigilant.

I am speaking in generalities here, and things will, of course, go differently in different real cases. Even so, I think these general trends are likely, based on the way that these features of standard practice operate. I also suspect that, at a very high level of abstraction, something like these stories is actually true of the historical development of comparative psychology. I don't mean that these dynamics were *the* core drivers of history, but I think they are among the factors that explain the current alignment of the field. It's not entirely an accident that so much research in comparative psychology is organized by these four features together.

Even setting aside the specific dynamics, I hope it's clear that these four features support one another in powerful ways. As a result, I think of them together as being like water molecules in a droplet on a leaf; their mutual attraction creates a powerful kind of surface tension. Trying to simply remove one feature while leaving the others intact is akin to trying to remove a single quadrant of a water droplet. The rest of the water will automatically and instantly rush in and fill the space. We still end up with a water droplet. Perhaps it is now a little smaller, but the shape is the same, and we have not succeeded in actually removing any one 'piece.'

Because of this, the field needs systematic replacements that avoid this entire structure. To criticize each feature individually is not enough. We need to wipe the leaf dry and start with something new. There is not one way to do this, and perhaps there will never be one that fits together internally as well as the current standard practice. There likely will not be one that applies so directly across species and target capacity. Even so, holism and modesty, as suggested through the book, are a good general starting point.

Before I move on to more specific suggestions, I want to reiterate my optimism about the general direction of the field. I know that framing the book around *challenges* may still feel intrinsically critical or pessimistic. I also just now framed my positive suggestions against criticism of 'standard practice' in comparative psychology. But still, I do think the field is making progress, and I do think it will continue to do so. Indeed, I get the sense that much of the field is moving in a direction much like the one I suggest here. The goal of focusing on the challenges was to show that there is reason for optimism even when we confront the challenges head-on. The future of the science, as I see it, can be richer and more active than ever.

Over the course of the book, my arguments have implied various specific suggestions that apply at different stages of the scientific process, which I summarize here by moving through those stages. One might agree with some suggestions and not others; this would not bother me much. The overall approach is the most important thing here. These specific suggestions are simply predictions about which approaches are most likely to make the science more fruitful. The only way to be sure what works best is to try it and see.

8.2 Model Evaluation

The most significant suggestion is that individual behavioral experiments typically only provide small pieces of evidence about the cognitive mechanisms behind them. Experiments can provide decisive evidence about a particular *statistical* hypothesis, but not a *substantive* hypothesis. Since substantive hypotheses carry the claims of interest about cognition, these are the hypotheses of interest. Addressing a substantive hypothesis requires a second step, beyond the statistical result. This is summed up in the two-stage evidentialist approach presented in Chapter 1 (esp. Figure 1.4). Each individual experimental finding joins the pool of evidence, which includes other related findings as well as contributions of background theory that might bear on the capacity at hand. The actual evaluation of candidate models requires a holistic inference to the best explanation based on this collected evidence. In the context of research on consciousness, this means gathering distinct clusters of putative measures and working with those as we develop and revise theory.

All along, researchers should recognize the possibility of small preferences for one model over another, and the possibility of withholding judgment if the evidence is genuinely unclear. The fact that uncertainty is endemic may add wrinkles here: we should be careful not to fall into a pattern of simply throwing up our hands. Careful handling of uncertainty also involves willingness to state what one thinks is most likely, even when confidence is relatively low.

The new challenge that arises with this approach is gathering and aggregating evidence when each piece only means a little bit. Perhaps the most direct approach is to aggregate evidence as it comes in. We could perform this aggregation along loosely Bayesian or likelihoodist lines: as each new piece of evidence comes in, we can assess how strongly it tells in one or the other direction for a given substantive hypothesis, and update our credences in candidate models accordingly. Thus, we can be likelihoodist at this stage, even if we apply significance testing statistical methods (more below). Moreover, assessments of the strength of evidence will need to be articulated and argued for; they should not be smuggled along with the false certainty of a calculation. This is part of the reason I see the inference as qualitative; even if scaffolded by formal methods in Bayesianism, the actual assessments of probabilities have to come from somewhere, and, in most cases in comparative psychology, they will be based on assessments of plausibility and reasonableness. Unless and until we have much more precise models (more below), there is no real way around this. However, it helps if we can discuss these issues in the open.

204 CONCLUSION

Various ways of modeling evidence can also scaffold the interpretation of experimental results. I suggested two main approaches under the general heading of 'cognitive cartography' or mapping out the 'shape' of the capacity along some set of criteria. One is dimensional modeling. Dimensional models break the big question of whether some capacity is present into smaller questions about how sophisticated the actual capacity is along different dimensions. Another is associative modeling, as I have argued it should be interpreted. That is, associative models should be seen as highly abstract, partial descriptions of the process, which can describe many different kinds of process of varying complexity. Here, the 'shape' is in the patterns of inferential sequence and the flow of information as the process moves through sequences of representations and guides behavior as a result. So interpreted, associative models can mediate between cognitive models and data in both directions, by summarizing sequences of representational states suggested by the data or by making more precise predictions about sequences of representational states from the model.

It is an open question which of these strategies, or others I have not considered, might be most fruitful. Most likely, their utility will vary based on the context of research—the questions being asked and the methods applied in answering them. Even so, the space of options is worth exploring, looking for ways of scaffolding model evaluation using individual pieces of evidence that are limited and partial. Most likely, these will be based in various ways of breaking bigger questions into something smaller.

8.3 Statistical Methods

I have resisted making claims about statistical methods because these can vary so much based on the kind of phenomenon one studies and data one gathers. However, I will note that the strategy I advocate includes null hypothesis significance testing (NHST) and even a possible defense against some criticisms of it (Dacey 2023). I've just listed NHST as a feature of standard practice that contributes to its troubles. Others have advocated for the abandonment of NHST for a variety of related reasons (e.g., Wagenmakers et al. 2008). So one might wonder why I retain it. In short, I think the two-stage evaluation of evidence just described largely alleviates these worries.

One common response to worries like the ones I have is to build more substance into the statistical methods. This is what Bayesian statistics do, as they require the choice of marginal likelihoods (the probability that a given result would be found if each hypothesis were true). Authors including Yarkoni

(2022) and Farrar, Voudouris, and Clayton (2021) similarly argue that data itself should be modeled with more sources of variance in order to make a given statistical result more generalizable. The shared idea here is to 'build up' the statistical decision-making process so that it can actually do the work that it is so often asked to do. But building up only makes sense if individual statistical results must do that work. My suggestion moves in the opposite direction, to 'pare down' individual statistical results.

The idea of paring down is to interpret results as thinly as possible ("there is some effect here"), and then separately evaluate what that result might mean for the models under consideration ("the model we accept should explain the effect"). I take this to avoid some of the problems with NHST. For example, Bayesian statisticians sometimes argue that NHST overweights the evidence that comes with 'positive' results and underweights the evidence that comes with 'negative' results (e.g., Rouder et al. 2009). If we pare down, the actual interpretation we apply to a given result (positive or negative) in the holistic inference to the best explanation will determine how much weight it is given. Generally, this reduces the weight granted positive results, but it also opens room for consideration of negative results. We don't need to change the statistical methods themselves; we just need to change the way we look at the results they give us.

The goal is that the results themselves can be widely agreed upon as a baseline because there are minimal contestable assumptions baked into the statistical methods. If we 'build up' by including more contestable assumptions in the statistical method, we shuttle disagreement into the statistical test. This risks producing statistical results that are not agreed upon because the assumptions built into the test are in dispute as well. Researchers working with different assumptions can produce distinct sets of evidence, and mutual engagement becomes difficult. If the experimental results themselves include as few contestable assumptions as possible (paring down), it is more likely that we can have agreement that *these* are the results we need to explain. The question is what to make of them, and how various candidate models might explain them. This discussion is hard but can take place in the open, and the experimental results can serve as common currency even between those who disagree on the theory.

This is a large issue that I cannot do true justice to here. And most likely different methods work for different purposes in different places. For now, I merely suggest that paring down be considered as an option, and that building up should not be the only option pursued by those who share the kinds of worry I've discussed here.

Regarding sample sizes: researchers have been attempting to, and should continue to, work on ways of increasing sample sizes. I think this is likely to continue to be an issue due to various practical limitations. And small sample experiments still have value, so they do not necessarily need to be entirely eliminated. They should be interpreted in ways that treat them in some ways as standard experiments and in other ways as case studies or anecdotes. More concretely, in Chapter 6 I also suggested registered reports. This is a process by which journals accept papers based only on methods, before the experiments are done. This practice could reduce the worry that publication bias in favor of positive results is systematically selecting for flukes that may be more likely in small samples. Worries about generalizability and replicability for small sample studies will remain, but they are reduced if we treat them carefully. When any one experiment is less important, a single failure to replicate is also less of a blow. Generalizability is possible as long as it is careful, sufficiently scaffolded, and partial, as one might generalize conclusions from a case study.

8.4 Experimental Design

Here again, the actual implications of my approach vary by context. Nonetheless, there are a couple of high-level suggestions for the design of experiments. The first is to move away from success testing. That is, to run more experiments in which it is not the case that finding some effect indicates a 'success' on the part of the animal. More experiments should attempt to show effects that are failures or are simply neutral with respect to whatever normative standards might be applied. For example, we might test for the presence of a reasoning bias that leads to a particular kind of error (as in the heuristics and biases literature using human participants; e.g., Kahneman et al. 1982). This prevents the accidental recreation of Morgan's Canon by aggregating studies that all give the privileged status of the null to statistical hypotheses better predicted by models describing simpler processes. Taylor et al. (2022) suggest this as a move to 'signature testing' over 'success testing.' I agree in substance, though I am wary of the 'signatures' terminology, which might imply that a specific test could tell between two hypotheses, if only we could find the right signature. Instead, I think the differences are more likely to emerge over patterns of results in lots of experiments.

In cases in which researchers are concerned with ecological validity, they should look for ways to design anchoring experiments that can connect the experimental behaviors back to natural behaviors. This, coupled with the kind of systematic exploration described in Chapter 5 can provide some protection from problems associated with the artificiality of tasks and contexts as well as the developmental background of the animals. Also, variations away from anchoring experiments could be another way to design tests that aren't tests of success: a researcher can manipulate a new variable that has not been tested in the laboratory context yet, but might tell us more about how the animal is doing what it's doing.

Finally, I do think it should be open that researchers sometimes run experiments that are not, themselves, intended to test between models. Experiments can simply be intended to systematically vary parameters relevant to performance on some task, and see what happens. This way, we learn a lot about the capacity responsible but need not aim to provide a crucial test between hypotheses. I think this can be done without recreating Allen's (2014) "trophy hunting" if we move away from success testing and interpret individual experiments as described here. As a general matter, because individual experiments do not decide between substantive hypotheses, the ability of a single experiment to decide between substantive hypotheses should not be the only criterion with which we decide which potential experiments to do.

8.5 Building Models

At the most general level, comparative psychology should work towards producing more precise, perhaps mathematical, models (Allen 2014). The particular strategy I advocated here is to interpret associative models as partial descriptions, which frees them up to be used in contexts which their current interpretation doesn't allow. Partial models can make specific testable predictions, even if they do not tell us the 'true nature' of the process involved. I have developed this view for associative models. However, I suspect there are others for which it can be used. Here, again, is a question of utility; whichever models describe and predict the things that matter most for one's research questions are the ones to apply.

In general, though, we should expect that no single model can fully capture a psychological process. It is likely that many models will be involved, at different levels, describing behavior in different kinds of tasks or under different conditions. The multi-model strategy can help build up understanding piecemeal, and solve some problems we have with the lack of precise modeling at the same time.

208 CONCLUSION

Development of models (of cognitive processes and of consciousness) should be responsive to new evidence as it comes in. The kinds of partial models I have described, I suggest, provide ways to systematically incorporate results without creating models that are perniciously ad hoc. The process of refining and re-evaluating models is part of a holistic inference to the best explanation, so changes need to be argued for based on how they fit the current state of overall evidence and background theory (and not merely as a response to the single newest experiment).

8.6 Parting Thoughts

These suggestions are summarized here at a high level of abstraction and can be made more precise in various different ways. As such, empirical researchers working in different areas would retain substantial control; there is a lot of expertise in comparative psychology, and I would not claim to be able to give specific advice.

And while these suggestions touch on different aspects of the science, I hope it is evident that they fit together and replace most of the four features of standard practice. Modesty about experimental results goes with modesty about parsimony claims, because the strength of preexisting parsimony claims sets the bar that experimental evidence must clear (and vice versa). As a result, we abandon the default framing (Chapter 1) and need new ways to structure the role that evidence plays. One promising option is provided by more modest interpretation of associative models. This, in turn, breaks apart the associative/cognitive divide. Requiring that we look across experiments to evaluate hypotheses presents further opportunities. We can get creative in experimental design to move away from success testing. We can organize experiments in ways that allow for anchoring experiments, or that allow for multiple epistemic clusters of measures of consciousness to be pursued together. It makes room to reinterpret small sample studies in ways that match other practices, such as case studies. Of the four features above, Morgan's Canon and null hypothesis significance testing can be retained if they are reinterpreted. We just need to interpret them modestly. Namely, Morgan's Canon can be interpreted as a kind of empirical parsimony (Chapter 1; Dacey 2016a), and NHST results can be 'pared down' in their interpretation, as just described. If we do this, they won't have the same effects.

This is the beginning of a sketch of a coherent replacement for standard practice. There is a lot of room for things to be filled in, and a lot of room for disagreement on particulars. The effect I most hope this book to have, though, is to help open up new paths to explore. The future of the science of animal minds is bright. There is a lot of work to do here, and even if the suggestions from this book are pointing in the right direction, there is a long way to go. But that is not a new problem; we have known it all along. The scientific study of animal minds is difficult.

References

Alberghina, D., Stevens, J. R., Silver, Z. A., Sexton, C. L., Rothkoff, L., Ravid-Schurr, D., . . . and Buchsbaum, D. (2023). "ManyDogs project: A big team science approach to investigating canine behavior and cognition." *Comparative Cognition & Behavior Reviews*, 18, 59–77.

Alem, S., Perry, C. J., Zhu, X., Loukola, O. J., Ingraham, T., Søvik, E., and Chittka, L. (2016). "Associative mechanisms allow for social learning and cultural transmission of string pulling in an insect." *PLOS Biology*, 14(10), e1002564.

Allen, C. (2014). "Models, mechanisms, and animal minds." *Southern Journal of Philosophy*, 52, 75–97.

Allen, C., and Bekoff, M. (1999). *Species of Mind: The Philosophy and Biology of Cognitive Ethology*. MIT Press.

Allen, C., and Trestman, M. (2020). "Animal consciousness." In *The Stanford Encyclopedia of Philosophy* (Winter 2020 Edition), ed. Edward N. Zalta, https://plato.stanford.edu/archives/win2020/entries/consciousness-animal/.

Alupay, J. S., Hadjisolomou, S. P., and Crook, R. J. (2014). "Arm injury produces long-term behavioral and neural hypersensitivity in octopus." *Neuroscience Letters*, 558, 137–42.

Anderson, R. C., and Mather, J. A. (2010). "It's all in the cues: octopuses (Enteroctopus dofleini) learn to open jars." *Ferrantia*, 59, 8–13.

Anderson, M. L., Richardson, M. J., and Chemero, A. (2012). "Eroding the boundaries of cognition: Implications of embodiment 1." *Topics in Cognitive Science*, 4(4), 717–30.

Andrews, K. (2012). *Do Apes Read Minds? Toward a New Folk Psychology*. MIT Press.

Andrews, K. (2014). *The Animal Mind: An Introduction to the Philosophy of Animal Cognition*. Routledge.

Andrews, K. (2020). *How to Study Animal Minds*. Cambridge University Press.

Andrews, K., and Huss, B. (2014). "Anthropomorphism, anthropectomy, and the null hypothesis." *Biology & Philosophy*, 29(5), 711–29.

Ankeny, R. (2011). *Using Cases to Establish Novel Diagnoses: Creating Generic Facts by Making Particular Facts Travel Together*. Cambridge University Press.

Araújo, D., Davids, K., and Passos, P. (2007). "Ecological validity, representative design, and correspondence between experimental task constraints and behavioral setting: Comment on." *Ecological Psychology*, 19(1), 69–78.

Arico, A., Fiala, B., Goldberg, R. F., and Nichols, S. (2011). "The folk psychology of consciousness." *Mind & Language*, 26(3), 327–52.

Baars, B. J. (1993). *A Cognitive Theory of Consciousness*. Cambridge University Press.

Baars, B. J. (2005). "Global workspace theory of consciousness: Toward a cognitive neuroscience of human experience." *Progress in Brain Research*, 150, 45–53.

Baars, B. J. (2017). "The global workspace theory of consciousness: Predictions and results." In S. Schneider and M. Velmans (eds.), *The Blackwell Companion to Consciousness*, 2nd ed., 227–42. Wiley-Blackwell.

Barrett, J. L., and Keil, F. C. (1996). "Conceptualizing a nonnatural entity: Anthropomorphism in God concepts." *Cognitive Psychology*, 31(3), 219–47.

212 REFERENCES

Barrett, L. F. (2017). *How Emotions are Made: The Secret Life of the Brain*. Pan Macmillan.

Barron, A. B., and Klein, C. (2016). "What insects can tell us about the origins of consciousness." *Proceedings of the National Academy of Sciences*, 113(18), 4900–8.

Bateson, M., Desire, S., Gartside, S. E., and Wright, G. A. (2011). "Agitated honeybees exhibit pessimistic cognitive biases." *Current Biology*, 21(12), 1070–3.

Bausman, W., and Halina, M. (2018). "Not null enough: pseudo-null hypotheses in community ecology and comparative psychology." *Biology & Philosophy*, 33(3–4), article no. 30.

Bayne, T., and Shea, N. (2020). "Consciousness, concepts, and natural kinds." *Philosophical Topics*, 48(1), 65–84.

Bechtel, W., and Abrahamsen, A. (1991). *Connectionism and the Mind: An Introduction to Parallel Processing in Networks*. Basil Blackwell.

Bekoff, M. (2000). "Animal emotions: exploring passionate natures." *BioScience*, 50(10), 861–70.

Bekoff, M., and Allen, C. (1997). "Cognitive ethology: Slayers, skeptics, and proponents." In R. W. Mitchell, N. S. Thompson, and H. L. Miles (eds.), *Anthropomorphism, Anecdotes, and Animals*, 313–34. State University of New York Press.

Beran, M. (2018). "Replication and pre-registration in comparative psychology." *International Journal of Comparative Psychology*, 31.

Beran, M. (2020). "Editorial: The value and status of replications in animal behavior and cognition research." *Animal Behavior and Cognition* 7(1): i–iii.

Birch, J. (2017). "Animal sentience and the precautionary principle." *Animal Sentience*, 2(16), 1.

Birch, J. (2022) "The search for invertebrate consciousness." *Noûs*, 56, 133–53.

Birch, J., Broom, D. M., Browning, H., Crump, A., Ginsburg, S., Halina, M., . . . and Zacks, O. (2022). "How should we study animal consciousness scientifically?" *Journal of Consciousness Studies*, 29(3–4).

Birch, J., Burn, C. Schnell, A., Browning, H., and Crump, A. (2021). *Review of the Evidence of Sentience in Cephalopod Molluscs and Decapod Crustaceans*. LSE Consulting. London School of Economics and Political Science.

Birch, J., Ginsburg, S., and Jablonka, E. (2020). "Unlimited associative learning and the origins of consciousness: a primer and some predictions." *Biology & Philosophy*, 35(6), 1–23.

Birch, J., Schnell, A. K., and Clayton, N. S. (2020). "Dimensions of animal consciousness." *Trends in Cognitive Sciences*, 24(10), 789–801.

Blair, R. J. R. (2012). "Considering anger from a cognitive neuroscience perspective." *Wiley Interdisciplinary Reviews: Cognitive Science*, 3(1), 65–74.

Blaisdell, A. P., Sawa, K., Leising, K. J., and Waldmann M. R. (2006). "Causal reasoning in rats." *Science*, 311, 1020–2.

Blaisdell, A. P., and Waldmann, M. R. (2012) "Rational rats: Causal inference and representation." In E. R. Wasserman and T. R. Zentall (eds.), *The Oxford Handbook of Comparative Cognition*. Oxford University Press.

Blanchard, D. C., and Blanchard, R. J. (1984) "Affect and aggression: an animal model applied to human behavior." In R. J. Blanchard and D. C. Blanchard (eds.), *Advances in the Study of Aggression*, vol. 1, 1–62.

Blanchard, D. C., and Blanchard, R. J. (1988) "Ethoexperimental approaches to the biology of emotion." *Annual Review of Psychology*, 39, 43–68.

Blanchard, D. C., and Blanchard, R. J. (2003) "What can animal aggression research tell us about human aggression?" *Hormones and Behavior,* 44(3), 171–7.

Block, N. (1995). "On a confusion about a function of consciousness." *Behavioral and Brain Sciences*, 18(2), 227–47.

REFERENCES   213

Block, N. (2007). "Consciousness, accessibility, and the mesh between psychology and neuroscience." *Behavioral and Brain Sciences*, 30(5–6), 481–99.

Boone, W., and Piccinini, G. (2016). "The cognitive neuroscience revolution." *Synthese*, 193(5), 1509–34.

Borrelli, L., and Fiorito, G. (2008). "Behavioral analysis of learning and memory in cephalopods." In R. Menzel (ed.), *Learning Theory and Behavior* 605–28. Elsevier.

Boyd, R. (1991). "Realism, anti-foundationalism and the enthusiasm for natural kinds." *Philosophical Studies*, 61(1/2), 127–48.

Boyle, A. (2019). "Mapping the minds of others." *Review of Philosophy and Psychology*, 10(4), 747–67.

Boyle, A. (2021). "Replication, uncertainty and progress in comparative cognition." *Animal Behavior and Cognition*, 8(2), 296–304.

Brigandt, I. (2003). "Homology in comparative, molecular, and evolutionary developmental biology: the radiation of a concept." *Journal of Experimental Zoology Part B: Molecular and Developmental Evolution*, 299(1), 9–17.

Browning, H., and Veit, W. (2020). "The measurement problem of consciousness." *Philosophical Topics*, 48(1), 85–108.

Brunswik, E. (1943). "Organismic achievement and environmental probability." *Psychological Review*, 50(3), 255.

Bub, J., and Bub, D. (1988). "On the methodology of single-case studies in cognitive neuropsychology." *Cognitive Neuropsychology*, 5, 563–82.

Bublitz, A., Dehnhardt, G., and Hanke, F. D. (2021). "Reversal of a spatial discrimination task in the common octopus (Octopus vulgaris)." *Frontiers in Behavioral Neuroscience*, 15, 614523.

Bublitz, A., Weinhold, S. R., Strobel, S., Dehnhardt, G., and Hanke, F. D. (2017). "Reconsideration of serial visual reversal learning in octopus (Octopus vulgaris) from a methodological perspective." *Frontiers in Physiology*, 8, 54.

Buckner, C. (2011). "Two approaches to the distinction between cognition and 'mere association.'" *International Journal of Comparative Psychology*, 24(4), 314–48.

Buckner, C. (2013). "Morgan's Canon, meet Hume's Dictum: avoiding anthropofabulation in cross-species comparisons." *Biology & Philosophy*, 28(5), 853–71.

Buckner, C. (2014). "The semantic problem(s) with research on animal mind-reading." *Mind & Language*, 29(5), 566–89.

Buckner, C. (2017). "Understanding associative and cognitive explanations in comparative psychology." In K. Andrews and J. Beck (eds.), *The Routledge Handbook of Philosophy of Animal Minds,* 409–18. Routledge.

Buckner, C. (2019). "Rational inference: The lowest bounds." *Philosophy and Phenomenological Research*, 98(3), 697–724.

Buller, D. J. (2006). *Adapting Minds: Evolutionary Psychology and the Persistent Quest for Human Nature.* MIT Press.

Bullock, T. H. (1984). "Ongoing compound field potentials from octopus brain are labile and vertebrate-like." *Electroencephalography and Clinical Neurophysiology*, 57(5), 473–83.

Bullock, T. H., and Budelmann, B. U. (1991). "Sensory evoked potentials in unanesthetized unrestrained cuttlefish: a new preparation for brain physiology in cephalopods." *Journal of Comparative Physiology A*, 168, 141–50.

Burghardt, G. M. (1991). "Cognitive ethology and critical anthropomorphism: A snake with two heads and hognose snakes that play dead." In C. A. Ristau (ed.), *Cognitive*

214 REFERENCES

Ethology: The Minds of Other Animals: Essays in Honor of Donald R. Griffin, 53–90. Lawrence Erlbaum Associates.

Burghardt, G. M. (2007). "Critical anthropomorphism, uncritical anthropocentrism, and naïve nominalism." *Comparative Cognition & Behavior Reviews*, 2.

Butterfill, S. A., and Apperly, I. A. (2013). "How to construct a minimal theory of mind." *Mind & Language*, 28(5), 606–37.

Campbell, D. T. (1975). " 'Degrees of Freedom' and the case study." *Comparative Political Studies*, 8(2): 178–93.

Canteras, N. S. (2002). "The medial hypothalamic defensive system: hodological organization and functional implications." *Pharmacology Biochemistry & Behavior*, 71(3):481–91.

Caporael, L. R., and Heyes, C. M. (1997). "Why anthropomorphize? Folk psychology and other stories." In R. W. Mitchell, N. S Thompson, and H. L. Miles (eds.), *Anthropomorphism, Anecdotes, and Animals*, 59–73. State University of New York Press.

Caramazza, A. (1986). "On drawing inferences about the structure of normal cognitive systems from the analysis of patterns of impaired performance: The case for single-patient studies." *Brain and Cognition*, 5, 41–66.

Carroll, N. C., and Young, A. W. (2005). "Priming of emotion recognition." *Quarterly Journal of Experimental Psychology Section A*, 58(7), 1173–97.

Carruthers, P. (2018). "Comparative psychology without consciousness." *Consciousness and Cognition*, 63, 47–60.

Cartwright, N. (2004). "Causation: One word, many things." *Philosophy of Science*, 71(5), 805–19.

Chalmers, D. J. (1995). "Facing up to the problem of consciousness." *Journal of Consciousness Studies*, 2(3), 200–19.

Chalmers, D. J. (1996). *The Conscious Mind: In Search of a Fundamental Theory*. Oxford University Press.

Chang, H. (2004). *Inventing Temperature: Measurement and Scientific Progress*. Oxford University Press.

Chemero, A., and Silberstein, M. (2008). "After the philosophy of mind: Replacing scholasticism with science." *Philosophy of Science*, 75(1), 1–27.

Cheney, D. L., and Seyfarth, R. M. (1980). "Vocal recognition in free-ranging vervet monkeys." *Animal Behaviour*, 28(2), 362–7.

Chittka, L. (2017). "Bee cognition." *Current Biology*, 27(19), R1049–53.

Chomsky, N. (1957). *Syntactic Structures*. Mouton & Co.

Chugunova, M., and Sele, D. (2020). *We and It: An Interdisciplinary Review of the Experimental Evidence on How Humans Interact with Machines*. SSRN, https://doi.org/10.2139/ssrn.3692293.

Churchland, P. M. (1981). "Eliminative materialism and propositional attitudes." *Journal of Philosophy*, 78(2), 67–90.

Clark, A. (1993). *Associative Engines: Connectionism, Concepts, and Representational Change*. MIT Press.

Clark, A. (2015). "Radical predictive processing." *Southern Journal of Philosophy*, 53, 3–27.

Colombo, M., Elkin, L., and Hartmann, S. (2021). "Being realist about Bayes, and the predictive processing theory of mind." *British Journal for the Philosophy of Science*, 72(1):185–220.

Colombo, M., and Seriès, P. (2012). "Bayes in the brain—On Bayesian modelling in neuroscience." *British Journal for the Philosophy of Science*, 63(3).

Cosmides, L., and Tooby, J. (1992). "Cognitive adaptations for social exchange." In J. Barkow, L. Cosmides, and J. Tooby (eds.), *The Adapted Mind*, 163–228. Oxford University Press.

REFERENCES   215

Cosmides, L., and Tooby, J. (1994). "Origins of domain specificity: The evolution of functional organization." In L. A. Hirschfeld and S. Gelman (eds.), *Mapping the Mind: Domain Specificity in Cognition and Culture*, 85–116. Cambridge University Press.

Cosmides, L., and Tooby, J. (2005). "Neurocognitive adaptations designed for social exchange." In D. Buss (ed.), *The Handbook of Evolutionary Psychology*, 3–87. Wiley.

Craver, C. F. (2007). *Explaining the Brain: Mechanisms and the Mosaic Unity of Neuroscience.* Clarendon Press.

Crook, R. J. (2021). "Behavioral and neurophysiological evidence suggests affective pain experience in octopus." *Iscience*, 24(3), 102229.

Crook, R. J., and Walters, E. T. (2011). "Nociceptive behavior and physiology of molluscs: animal welfare implications." *Ilar Journal*, 52(2), 185–95.

Cross, E. S., Riddoch, K. A., Pratts, J., Titone, S., Chaudhury, B., and Hortensius, R. (2019). "A neurocognitive investigation of the impact of socializing with a robot on empathy for pain." *Philosophical Transactions of the Royal Society B*, 374(1771), 20180034.

Crump, A., and Birch, J. (2021). "Separating conscious and unconscious perception in animals." *Learning & Behavior*, 49(4), 347–8.

Cummins, R. (2002) " 'How does it work?' vs. 'What are the laws?' Two conceptions of psychological explanation." In F. Keil and R. Wilson (eds.), *Explanation and Cognition,* 117–45. MIT Press.

Currie, A. (2021). *Comparative Thinking in Biology.* Cambridge University Press.

Dacey, M. (2015). "Associationism without associative links: Thomas Brown and the associationist project." *Studies in History and Philosophy of Science Part A*, 54, 31–40.

Dacey, M. (2016a). "The varieties of parsimony in psychology." *Mind & Language*, 31, 414–37.

Dacey, M. (2016b). "Rethinking associations in psychology." *Synthese*, 193(12), 3763–86.

Dacey, M. (2017a). "Anthropomorphism as cognitive bias." *Philosophy of Science*, 84(5), 1152–64.

Dacey, M. (2017b). "A new view of association and associative models." In K. Andrews and J. Beck (eds.), *The Routledge Handbook of Philosophy of Animal Minds*, 419–26. Routledge.

Dacey, M. (2019a). "Simplicity and the meaning of mental association." *Erkenntnis*, 84(6), 1207–28.

Dacey, M. (2019b). "Association and the mechanisms of priming." *Journal of Cognitive Science*, 20(3), 281–321.

Dacey, M. (2020). "Associationism in the philosophy of mind." *Internet Encyclopedia of Philosophy*, https://iep.utm.edu/associationism-in-philosophy-of-mind/.

Dacey, M. (2022a). "Separate substantive from statistical hypotheses and treat them differently." *Behavioral and Brain Sciences*, 45.

Dacey, M. (2022b). "The shared project, but divergent views, of the Empiricist associationists." *Philosophical Psychology*, 37(4), 759–81.

Dacey, M. (2023). "Evidence in default: Rejecting default models of animal minds." *British Journal for the Philosophy of Science*, 74(2), 291–312.

Dacey, M., and Coane, J. H. (2023). "Implicit measures of anthropomorphism: affective priming and recognition of apparent animal emotions." *Frontiers in Psychology*, 14, 1149444.

Danks, D. (2014). *Unifying the Mind: Cognitive Representations as Graphical Models.* MIT Press.

Darwin, C. (1871). *The Descent of Man.* D. Appleton & Co.

Darwin, C. (1877). *The Effects of Cross and Self Fertilisation in the Vegetable Kingdom.* D. Appleton & Co.

216 REFERENCES

Deecke, V. B. (2006). "Studying marine mammal cognition in the wild: a review of four decades of playback experiments." *Aquatic Mammals*, 32(4), 461–82.

de Geus, E. J., Wright, M. J., Martin, N. G., and Boomsma, D. I. (2001). "Genetics of brain function and cognition." *Behavior Genetics*, 31(6), 489–95.

Dehaene, S., and Changeux, J. P. (2011). "Experimental and theoretical approaches to conscious processing." *Neuron*, 70(2), 200–27.

Del Cul, A., Baillet, S., and Dehaene, S. (2007). "Brain dynamics underlying the nonlinear threshold for access to consciousness." *PLOS Biology*, 5(10), e260.

Dennett, D. C. (1978). "Beliefs about beliefs." *Behavioral and Brain Sciences*, 1(4), 568–70.

Denniston, J. C., Savastano, H. I., Blaisdell, A. P., and Miller, R. R. (2003). "Cue competition as a retrieval deficit." *Learning & Motivation*, 34, 1–31.

de Waal, F. B. (1988). "The communicative repertoire of captive bonobos (Pan paniscus), compared to that of chimpanzees." *Behaviour*, 106(3–4), 183–251.

de Waal, F. B. (1991). "Complementary methods and convergent evidence in the study of primate social cognition." *Behaviour*, 118(3), 297–320.

de Waal, F. B. (1999). "Anthropomorphism and anthropodenial: consistency in our thinking about humans and other animals." *Philosophical Topics*, 27(1), 255–80.

de Waal, F. B. (2001). "The inevitability of evolutionary psychology and the limitations of adaptationism: Lessons from the other primates." *International Journal of Comparative Psychology*, 14(1).

de Waal, F. B., and Ferrari, P. F. (2010). "Towards a bottom-up perspective on animal and human cognition." *Trends in Cognitive Sciences*, 14(5), 201–7.

Diamond, C. (2006). "The difficulty of reality and the difficulty of philosophy." In A. Crary and S. Shieh (eds.), *Reading Cavell*, 108–28. Routledge.

Dickinson, A,. and Burke, J. (1996). "Within-compound associations mediate the retrospective revaluation of causality judgements." *Quarterly Journal of Experimental Psychology*, 49B, 60–80.

Dickinson, A. (2012). "Associative learning and animal cognition." *Philosophical Transactions of the Royal Society B: Biological Sciences*, 367(1603), 2733–42.

Diener, E., Northcott, R., Zyphur, M. J., and West, S. G. (2022). "Beyond experiments." *Perspectives on Psychological Science*, 17 (4).

Doris, J. M. (2015). *Talking to Our Selves: Reflection, Ignorance, and Agency.* Oxford University Press.

Douglas, H. E. (2000). "Inductive risk and values in science." *Philosophy of Science*, 67, 559–79

Dwyer, D. M., Starns, J., and Honey, R. C. (2009). "'Causal reasoning' in rats: A reappraisal." *Journal of Experimental Psychology: Animal Behavior Processes*, 35(4), 578.

Dyer, F. C. (2002). "The biology of the dance language." *Annual Review of Entomology*, 47, 917–49.

Ebersole, C. R., Atherton, O. E., Belanger, A. L., Skulborstad, H. M., Allen, J. M., Banks, J. B., . . . and Nosek, B. A. (2016). "Many Labs 3: Evaluating participant pool quality across the academic semester via replication." *Journal of Experimental Social Psychology*, 67, 68–82.

Eddy, T. J., Gallup, G. G., and Povinelli, D. J. (1993). "Attribution of cognitive states to animals: Anthropomorphism in comparative perspective." *Journal of Social Issues*, 49(1), 87–101.

Edelman, D. B., Baars, B. J., and Seth, A. K. (2005). "Identifying hallmarks of consciousness in non-mammalian species." *Consciousness and Cognition*, 14(1), 169–87.

Ekman, P., and Oster, H. (1979). "Facial expressions of emotion." *Annual Review of Psychology*, 30(1), 527–54.

REFERENCES 217

Epley, N., Akalis, S., Waytz, A., and Cacioppo, J. T. (2008). "Creating social connection through inferential reproduction: Loneliness and perceived agency in gadgets, gods, and greyhounds." *Psychological Science*, 19(2), 114–20.

Epley, N., Waytz, A., and Cacioppo, J. T. (2007). "On seeing human: a three-factor theory of anthropomorphism." *Psychological Review*, 114(4), 864.

Ereshefsky, M. (2007). "Psychological categories as homologies: lessons from ethology." *Biology & Philosophy*, 22(5), 659–74.

Ereshefsky, M. (2009). "Homology: integrating phylogeny and development." *Biological Theory*, 4(3), 225–9.

Ereshefsky, M. (2012). "Homology thinking." *Biology & Philosophy*, 27(3), 381–400.

Farrar, B. G., Boeckle, M., and Clayton, N. S. (2020). "Replications in comparative cognition: What should we expect and how can we improve?" *Animal Behavior and Cognition*, 7(1), 1–22, doi: https://doi.org/10.26451/abc.07.01.02.2020.

Farrar, B. G., Voudouris, K., and Clayton, N. S. (2021). "Replications, comparisons, sampling and the problem of representativeness in animal cognition research." *Animal Behavior and Cognition*, 8(2), 273.

Finn, J. K., Tregenza, T., and Norman, M. D. (2009). "Defensive tool use in a coconut-carrying octopus." *Current Biology*, 19(23), R1069–70.

Fiorito, G., and Scotto, P. (1992). "Observational learning in Octopus vulgaris." *Science*, 256(5056), 545–7.

Fitzpatrick, S. (2008). "Doing away with Morgan's Canon." *Mind & Language*, 23(2), 224–46.

Fitzpatrick, S. (2017). "Against Morgan's Canon." In K. Andrews and J. Beck (eds.), *The Routledge Handbook of Philosophy of Animal Minds*, 437–47. Routledge.

Forrester, J. (1996). "If p, then what? thinking in cases." *History of the Human Sciences*, 9(3), 1–25.

Free, J. B. (1963). "The flower constancy of honeybees." *The Journal of Animal Ecology*, 1, 119–31. https://www.jstor.org/stable/2521?origin=crossref

Friston, K. (2012). "The history of the future of the Bayesian brain." *NeuroImage*, 62(2), 1230–3.

Galef, B. (1996). "Historical origins: the making of a science." In L. Houck and L. Drickamer (eds.), *Foundations of Animal Behavior: Classic Papers with Commentaries*, 5–12. Chicago University Press.

Gannon, L. (2002). "A critique of evolutionary psychology." *Psychology, Evolution & Gender*, 4(2): 173–218.

Giancola-Detmering, S., and Crook, R. J. (2024). "A judgment bias task reveals stress produces a negative affective state in cuttlefish." bioRxiv, https://www.biorxiv.org/content/10.1101/2024.04.15.589613v1.

Ginsburg, S., and Jablonka, E. (2019). *The Evolution of the Sensitive Soul: Learning and the Origins of Consciousness*. MIT Press.

Glymour, C. (1994). "On the methods of cognitive neuropsychology." *British Journal for the Philosophy of Science*, 45(3), 815–35.

Glymour, C. (2001). *The Mind's Arrows: Bayes Nets and Graphical Causal Models in Psychology*. MIT Press.

Godfrey-Smith, P. (1994). "Of nulls and norms." *PSA: Proceedings of the Biennial Meeting of the Philosophy of Science Association*, 1994(1), 280–90.

Godfrey-Smith, P. (2016). *Other Minds: The Octopus, the Sea, and the Deep Origins of Consciousness*. Farrar, Straus and Giroux.

218 REFERENCES

Godfrey-Smith, P. (2019). "Evolving across the explanatory gap." *Philosophy, Theory, and Practice in Biology*, 11(1).

Godfrey-Smith, P. (2020). *Metazoa: Animal Life and the Birth of the Mind*. Farrar, Straus and Giroux.

Gopnik A., Glymour C., Sobel D., Schulz L., Kushnir T., and Danks D. (2004). "A theory of causal learning in children: causal maps and Bayes nets." *Psychological Review*, 111, 1–31.

Gould, J. L., and Gould, C. G. (1982). "The insect mind: physics or metaphysics?". In D. R. Griffin (ed.), *Animal Mind—Human Mind*, 269–97. Springer.

Gould, S. J., and Lewontin, R. C. (1979). "The spandrels of San Marco and the Panglossian paradigm: A critique of the adaptationist programme." *Proceedings of the Royal Society B*, 205(1161).

Gray, K., Young, L., and Waytz, A. (2012). "Mind perception is the essence of morality." *Psychological Inquiry*, 23(2), 101–24.

Green, L., and Myerson, J. (2004). "A discounting framework for choice with delayed and probabilistic rewards." *Psychological Bulletin*, 130(5), 769.

Griffin, D. R. (1976). *The Question of Animal Awareness: Evolutionary Continuity of Mental Experience*. Rockefeller University Press.

Griffiths, P. E. (1997). *What Emotions Really Are*. University of Chicago Press.

Gruen, L. (2015). *Entangled Empathy: An Alternative Ethic for Our Relationships with Animals*. Lantern Books.

Gutnick, T., Neef, A., Cherninskyi, A., Ziadi-Künzli, F., Di Cosmo, A., Lipp, H. P., and Kuba, M. J. (2023). "Recording electrical activity from the brain of behaving octopus." *Current Biology*, 33(6), 1171–8.

Halina, M. (2015). "There is no special problem of mindreading in nonhuman animals." *Philosophy of Science*, 82(3), 473–90.

Halina, M. (2021). "Replications in comparative psychology." *Animal Behavior and Cognition*, 8(2), 263–72. https://doi.org/10.26451/abc.08.02.13.2021

Halina, M. (2024). *Animal Minds*. Cambridge University Press.

Halina, M. (forthcoming). *Kinds of Intelligence*. Cambridge University Press.

Halina, M., Harrison, D., Klein, C. (2022). "Evolutionary transition markers and the origins of consciousness." *Journal of Consciousness Studies*, 29, 3–4; 62–77.

Haller, J. (2013). "The neurobiology of abnormal manifestations of aggression—a review of hypothalamic mechanisms in cats, rodents, and humans." *Brain Research Bulletin*, 93, 97–109.

Hanus, D. (2016). "Causal reasoning versus associative learning: A useful dichotomy or a strawman battle in comparative psychology?" *Journal of Comparative Psychology*, 130(3), 241.

Hare, B., Call, J., Agnetta, B., and Tomasello, M. (2000). "Chimpanzees know what conspecifics do and do not see." *Animal Behaviour*, 59(4), 771–85.

Hare, B., Call, J., and Tomasello, M. (2001). "Do chimpanzees know what conspecifics know?" *Animal Behaviour*, 61(1), 139–51.

Harris, J., and Anthis, J. R. (2021). "The moral consideration of artificial entities: a literature review." *Science and Engineering Ethics*, 27(4), 1–95.

Heider, F., and Simmel, M. (1944). "An experimental study of apparent behavior." *American Journal of Psychology*, 243–59.

Heisenberg, M. (1998). "What do the mushroom bodies do for the insect brain? An introduction." *Learning & Memory*, 5(1), 1–10.

Heyes, C. (1993). "Anecdotes, training, trapping and triangulating: do animals attribute mental states?" *Animal Behaviour*, 46(1), 177–88.

Heyes, C.(1998). "Theory of mind in nonhuman primates." *Behavioral and Brain Sciences*, 21(1), 101–14.

Heyes, C. (2014). "Submentalizing: I am not really reading your mind." *Perspectives on Psychological Science*, 9(2), 131–43.

Heyes, C. (2015). "Animal mindreading: what's the problem?" *Psychonomic Bulletin & Review*, 22(2), 313–27.

Heyes, C. (2017). "Apes submentalise." *Trends in Cognitive Sciences*, 21(1), 1–2.

Hill, C. S. (2016). "Insects: Still looking like zombies." *Animal Sentience*, 1(9), 20.

Hochstein, E. (2016). "One mechanism, many models: A distributed theory of mechanistic explanation." *Synthese*, 193(5), 1387–1407.

Hofmann, M. M., Cheke, L. G., and Clayton, N. S. (2016). "Western scrub-jays (Aphelocoma californica) solve multiple-string problems by the spatial relation of string and reward." *Animal Cognition*, 19(6), 1103–14.

Horowitz, A. C., and Bekoff, M. (2007). "Naturalizing anthropomorphism: Behavioral prompts to our humanizing of animals." *Anthrozoös*, 20(1), 23–35.

Hortensius, R., Hekele, F., and Cross, E. S. (2018). "The perception of emotion in artificial agents." *IEEE Transactions on Cognitive and Developmental Systems*, 10(4), 852–64.

Howard, S. R., Avarguès-Weber, A., Garcia, J. E., Greentree, A. D., and Dyer, A. G. (2019). "Numerical cognition in honeybees enables addition and subtraction." *Science Advances*, 5(2), eaav0961.

Hume, D. (1748/1974). *Enquiries Concerning Human Understanding and Concerning the Principles of Morals,* ed. L. A. Selby-Bigge. Clarendon Press.

Hume, D. (1757/2007). "The natural history of religion." In Tom L. Beauchamp (ed.), *A Dissertation on the Passions. The Natural History of Religion. A Critical Edition*, 30–87. Clarendon Press.

Inoue, S., and Matsuzawa, T. (2007). "Working memory of numerals in chimpanzees." *Current Biology*, 17(23), R1004–5.

Irvine, E. (2012). *Consciousness as a Scientific Concept: A Philosophy of Science Perspective.* Springer.

Irvine, E. (2020). "Developing valid behavioral indicators of animal pain." *Philosophical Topics*, 48(1), 129–53.

Irvine, E. (2021a). "Assessing unlimited associative learning as a transition marker." *Biology & Philosophy*, 36(2), 1–5.

Irvine, E. (2021b). "Developing dark pessimism towards the justificatory role of introspective reports." *Erkenntnis*, 86(6), 1319–44.

Jacobs, I. F., and Osvath, M. (2015). "The string-pulling paradigm in comparative psychology." *Journal of Comparative Psychology*, 129(2), 89.

Jacovides, M. (2024). Hume on the Prospects for a Scientific Psychology. Unpublished manuscript.

Jékely, Gáspár, Peter Godfrey-Smith, and Fred Keijzer. (2021). "Reafference and the origin of the self in early nervous system evolution." *Philosophical Transactions of the Royal Society B,* 376(1821), 20190764.

Jozet-Alves, C., Bertin, M., and Clayton, N. S. (2013). "Evidence of episodic-like memory in cuttlefish." *Current Biology*, 23(23), R1033–5.

Kahneman, D., Slovic, S. P., Slovic, P., and Tversky, A. (eds). (1982). *Judgment under Uncertainty: Heuristics and Biases.* Cambridge University Press.

Kamin L. J. (1969). "Selection association and conditioning." In N. J. Mackintosh (ed.), *Fundamental Issues in Associative Learning*, 42–64. Dalhousie University Press.

220 REFERENCES

Kano, F., Krupenye, C., Hirata, S., Call, J., and Tomasello, M. (2017). "Submentalizing cannot explain belief-based action anticipation in apes." *Trends in Cognitive Sciences,* 21(9), 633–4.

Kano, F., Krupenye, C., Hirata, S., Tomonaga, M., and Call, J. (2019). "Great apes use self-experience to anticipate an agent's action in a false-belief test." *Proceedings of the National Academy of Sciences,* 116(42), 20904–9.

Karin-D'Arcy, M. (2005). "The modern role of Morgan's Canon in comparative psychology." *International Journal of Comparative Psychology,* 18(3).

Kennedy, J. S. (1992). *The New Anthropomorphism.* Cambridge University Press.

Key, Brian. (2016). "Why fish do not feel pain." *Animal Sentience* 1(3), 1.

Keyser, S. J., Miller, G. A., and Walker, E. (1978). "Cognitive science in 1978." *Unpublished report submitted to the Alfred P. Sloan Foundation, New York.*

Kihlstrom, J. F. (2021). "Ecological validity and 'ecological validity'." *Perspectives on Psychological Science,* 16(2), 466–71.

Klein, C., and Barron, A. B. (2016). "Insects have the capacity for subjective experience." *Animal Sentience,* 1(9), 1.

Klein, R. A., Ratliff, K. A., Vianello, M., Adams Jr, R. B., Bahník, Š., Bernstein, M. J., . . . and Nosek, B. A. (2014). "Investigating variation in replicability: A 'many labs' replication project." *Social Psychology,* 45(3), 142.

Knill, D. C., and Pouget, A. (2004). "The Bayesian brain: the role of uncertainty in neural coding and computation." *TRENDS in Neurosciences,* 27(12), 712–19.

Koch, C., Massimini, M., Boly, M., and Tononi, G. (2016). "Neural correlates of consciousness: progress and problems." *Nature Reviews Neuroscience,* 17(5), 307–21.

Koltermann, R. (1969). "Lern- und Vergessensprozesse bei der Honigbiene—aufgezeigt anhand von Duftdressuren." *Zeitschrift für vergleichende Physiologie,* 63(3), 310–34.

Krupenye, C., Kano, F., Hirata, S., Call, J., and Tomasello, M. (2016). "Great apes anticipate that other individuals will act according to false beliefs." *Science,* 354(6308), 110–14.

Kruschke, J. K. (2011). "Bayesian assessment of null values via parameter estimation and model comparison." *Perspectives on Psychological Science,* 6(3), 299–312.

Kuhn, T. (1962) *The Structure of Scientific Revolutions.* Chicago University Press.

Lambert, M., Farrar, B., Garcia-Pelegrin, E., Reber, S., and Miller, R. (2022). "ManyBirds: A multi-site collaborative Open Science approach to avian cognition and behavior research." *Animal Behavior and Cognition,* 9(1), 133–52.

Lamme, V. A. (2006). "Towards a true neural stance on consciousness." *Trends in Cognitive Sciences,* 10(11), 494–501.

Learmonth, M. J. (2020). "The matter of non-avian reptile sentience, and why it 'matters' to them: A conceptual, ethical and scientific review." *Animals,* 10(5), 901.

Leising, K. J., Wong, J., Waldmann, M. R., and Blaisdell, A. P. (2008). "The special status of actions in causal reasoning in rats." *Journal of Experimental Psychology: General,* 137(3), 514.

Leslie, S. J. (2008). "Generics: Cognition and acquisition." *Philosophical Review,* 117(1), 1–47.

Levine, J. (1983). "Materialism and qualia: The explanatory gap." *Pacific Philosophical Quarterly,* 64(4), 354–61.

Lewis, L. S., and Krupenye, C. (2021). "Theory of mind in nonhuman primates." In B. L. Schwartz and M. J. Beran (eds.), *Primate Cognitive Studies.* Cambridge University Press.

Li, Z., Terfurth, L., Woller, J. P., and Wiese, E. (2022). "Mind the machines: applying implicit measures of mind perception to social robotics." In *2022 17th ACM/IEEE International Conference on Human-Robot Interaction (HRI),* 236–45. IEEE.

REFERENCES 221

Lin, H., Werner, K. M., and Inzlicht, M. (2021). "Promises and perils of experimentation: The mutual-internal-validity problem." *Perspectives on Psychological Science*, 16(4), 854–63.

Lindquist, K. A., Wager, T. D., Kober, H., Bliss-Moreau, E., and Barrett, L. F. (2012). "The brain basis of emotion: a meta-analytic review." *Behavioral and Brain Sciences*, 35(3), 121.

Loukola, O. J., Solvi, C., Coscos, L., and Chittka, L. (2017). "Bumblebees show cognitive flexibility by improving on an observed complex behavior." *Science*, 355(6327), 833–6.

Love, A. C. (2007). "Functional homology and homology of function: Biological concepts and philosophical consequences." *Biology & Philosophy*, 22(5), 691–708.

Luce, R. D. (1995). "Four tensions concerning mathematical modeling in psychology." *Annual Review of Psychology*, 46, 1.

Lurz, R., and Andreassi, V. (2021). "Chimpanzees are mindreaders: On why they attribute seeing rather than sensing." *Philosophical Psychology*, 35(4), 1–28.

Lurz, R. W. (2009). "If chimpanzees are mindreaders, could behavioral science tell? Toward a solution of the logical problem." *Philosophical Psychology*, 22(3), 305–28.

Lurz, R. W. (2011). *Mindreading Animals: the Debate over What Animals Know about Other Minds*. MIT Press.

Lurz, R. W., Kanet, S., and Krachun, C. (2014). "Animal mindreading: A defense of optimistic agnosticism." *Mind & Language*, 29(4), 428–54.

MaBouDi, H., Barron, A. B., Li, S., Honkanen, M., Loukola, O. J., Peng, F., . . . and Solvi, C. (2021). "Non-numerical strategies used by bees to solve numerical cognition tasks." *Proceedings of the Royal Society B*, 288(1945), 20202711.

Machery, E. (2020). "What is a replication?" *Philosophy of Science*, 87(4), 545–67.

Machery, E. (2021). "The alpha war." *Review of Philosophy and Psychology*, 12(1), 75–99.

Mackintosh, N. J., and Mackintosh, J. (1963). "Reversal learning in Octopus vulgaris Lamarck with and without irrelevant cues." *Quarterly Journal of Experimental Psychology*, 15(4), 236–42.

Mameli, M. (2008). "On innateness: The clutter hypothesis and the cluster hypothesis." *Journal of Philosophy*, 105(12), 719–36.

ManyPrimates project members: Altschul, D. M., Beran, M. J., Bohn, M., Call, J., DeTroy, S., . . . and Flessert, M. (2019). "Establishing an infrastructure for collaboration in primate cognition research." *PLOS ONE*, 14(10).

Marchesi, S., Bossi, F., Ghiglino, D., De Tommaso, D., and Wykowska, A. (2021). "I am looking for your mind: Pupil dilation predicts individual differences in sensitivity to hints of human-likeness in robot behavior." *Frontiers in Robotics and AI*, 8, 167.

Marr, D. (1982). *Vision: A Computational Investigation into the Human Representation and Processing of Visual Information*. MIT Press.

Maselli, V., Al-Soudy, A. S., Buglione, M., Aria, M., Polese, G., and Di Cosmo, A. (2020). "Sensorial hierarchy in Octopus vulgaris's food choice: chemical vs. visual." *Animals*, 10(3), 457.

Mashour, G. A., Roelfsema, P., Changeux, J. P., and Dehaene, S. (2020). "Conscious processing and the global neuronal workspace hypothesis." *Neuron*, 105(5), 776–98.

Mather, J. (2008). "Cephalopod consciousness: behavioural evidence." *Consciousness and Cognition*, 17(1), 37–48.

Mather, J. (2021a). "Octopus consciousness: the role of perceptual richness." *NeuroSci*, 2(3), 276–90.

Mather, J. (2021b). "The case for octopus consciousness: Unity." *NeuroSci*, 2(4), 405–15.

Mather, J. (2022a). "The case for octopus consciousness: Temporality." *NeuroSci*, 3(2), 245–61.

222 REFERENCES

Mather, J. (2022b). "The case for octopus consciousness: Valence." *NeuroSci*, 3(4), 656–66.

Mather, J., and Anderson, R. C. (1999). "Exploration, play and habituation in octopuses (Octopus dofleini)." *Journal of Comparative Psychology*, 113(3), 333.

Matthen, M. (2007). "Defining vision: what homology thinking contributes." *Biology and Philosophy*, 22(5), 675–89.

McClelland, J. L., Rumelhart, D. E., and PDP Research Group. (1987). *Parallel Distributed Processing, Vol. 2: Explorations in the Microstructure of Cognition: Psychological and Biological Models*. MIT Press.

McCloskey, M., and Caramazza, A. (1988). "Theory and methodology in cognitive neuropsychology: A response to our critics." *Cognitive Neuropsychology*, 5, 583–623.

McLean, R. (1983). "Gastropod shells: a dynamic resource that helps shape benthic community structure." *Journal of Experimental Marine Biology and Ecology*, 69(2), 151–74.

McSweeney, F. K., and Murphy, E. S. (2014). *The Wiley Blackwell Handbook of Operant and Classical Conditioning*. John Wiley & Sons.

Meketa, I. (2014). "A critique of the principle of cognitive simplicity in comparative cognition." *Biology and Philosophy*, 29(5), 731–45.

Menne, I. M., and Lugrin, B. (2017). "In the face of emotion: a behavioral study on emotions towards a robot using the facial action coding system." In *Proceedings of the Companion of the 2017 ACM/IEEE International Conference on Human-Robot Interaction*, 205–6.

Merker, B. (2005). "The liabilities of mobility: A selection pressure for the transition to consciousness in animal evolution." *Consciousness and Cognition*, 14(1), 89–114.

Merker, B. (2007). "Consciousness without a cerebral cortex: A challenge for neuroscience and medicine." *Behavioral and Brain Sciences*, 30(1), 63–81.

Michel, M. (2019). "The mismeasure of consciousness: A problem of coordination for the perceptual awareness scale." *Philosophy of Science*, 86(5), 1239–49.

Mikhalevich, I. (2017). "Simplicity and cognitive models: avoiding old mistakes in new contexts." In K. Andrews and J. Beck (eds.), *The Routledge Handbook of Philosophy of Animal Minds*, 427–36. Routledge.

Mikhalevich, I., and Powell, R. (2020). "Minds without spines: Evolutionarily inclusive animal ethics." *Animal Sentience*, 29(1).

Mikhalevich, I., Powell, R., and Logan, C. (2017). "Is behavioural flexibility evidence of cognitive complexity? How evolution can inform comparative cognition." *Interface Focus*, 7(3), 20160121.

Millard, C., and Callard, F. (2020). "Thinking in, with, across, and beyond cases with John Forrester." *History of the Human Sciences*, 33(3–4), 3–14.

Miller, G. A. (2003). "The cognitive revolution: a historical perspective." *Trends in Cognitive Sciences*, 7(3), 141–4.

Miller, R. R., and Matute, H. (1996). "Biological significance in forward and backward blocking: resolution of a discrepancy between animal conditioning and human causal judgment." *Journal of Experimental Psychology: General*, 125(4), 370–86.

Mitchell, R. W. (1997). "Anthropomorphic anecdotalism as method." In, R. W. Mitchell, N. S. Thompson, and H. L. Miles (eds.), *Anthropomorphism, Anecdotes, and Animals*, 151–69. State University of New York Press.

Mitchell, R. W., Thompson, N. S. and Miles, H. L. (1997). *Anthropomorphism, Anecdotes, and Animals*. State University of New York Press.

Mook, D. G. (1983). "In defense of external invalidity." *American Psychologist*, 38(4), 379.

Morewedge, C. K., Preston, J., and Wegner, D. M. (2007). "Timescale bias in the attribution of mind." *Journal of Personality and Social Psychology*, 93(1), 1.

Morgan, C. L. (1894). *Introduction to Comparative Psychology*. Walter Scott.

REFERENCES 223

Morgan, M. S. (2012). "Case studies: One observation or many? Justification or discovery?" *Philosophy of Science*, 79(5), 667–77.

Morgan, M. S. (2020). " 'If p? Then what?' Thinking within, with, and from cases." *History of the Human Sciences*, 33(3–4), 198–217.

Mulhall, S. (2009). *The Wounded Animal: JM Coetzee and the Difficulty of Reality in Literature and Philosophy*. Princeton University Press.

Nagel, T. (1974). "What is it like to be a bat?" *Philosophical Review*, 83(4), 435–50.

Neely, J. H. (1977). "Semantic priming and retrieval from lexical memory: Roles of inhibitionless spreading activation and limited-capacity attention." *Journal of Experimental Psychology: General*, 106(3), 226.

Nelson, N. C. (2018). *Model Behavior: Animal Experiments, Complexity, and the Genetics of Psychiatric Disorders*. University of Chicago Press.

Nemati, F. (2022). "Anthropomorphism in the context of scientific discovery: Implications for comparative cognition." *Foundations of Science*, 1–19.

Núñez, R., Allen, M., Gao, R., Miller Rigoli, C., Relaford-Doyle, J., and Semenuks, A. (2019). "What happened to cognitive science?" *Nature Human Behaviour*, 3(8), 782–91.

O'Connell-Rodwell, C. E., Wood, J. D., Kinzley, C., Rodwell, T. C., Poole, J. H., and Puria, S. (2007). "Wild African elephants (Loxodonta africana) discriminate between familiar and unfamiliar conspecific seismic alarm calls." *Journal of the Acoustical Society of America*, 122(2), 823–30.

Open Science Collaboration (2012). "An open, large-scale, collaborative effort to estimate the reproducibility of psychological science." *Perspectives on Psychological Science*, 7(6), 657–60.

Palmer, T. D., and Ramsey, A. K. (2012). "The function of consciousness in multisensory integration." *Cognition*, 125(3), 353–64.

Panksepp, J. (1998). *Affective Neuroscience: The Foundations of Human and Animal Emotions*. Oxford University Press.

Panksepp, J. (2011). "The basic emotional circuits of mammalian brains: do animals have affective lives?" *Neuroscience & Biobehavioral Reviews*, 35(9), 1791–1804.

Panksepp, J., and Biven, L. (2012). *The Archaeology of Mind: Neuroevolutionary Origins of Human Emotions*. W.W. Norton & Co.

Panksepp, J., and Panksepp, J. B. (2000). "The seven sins of evolutionary psychology." *Evolution and Cognition*, 6(2), 108–31.

Panksepp, J., and Zellner, M. R. (2004). "Towards a neurobiologically based unified theory of aggression." *Revue internationale de psychologie sociale*, 17(2), 37–61.

Papineau, D., and Heyes, C. (2006). "Rational or associative? Imitation in Japanese quail." In S. Hurley and M. Nudds (eds.), *Rational Animals?*, 187–95. Oxford University Press.

Parr, L. A., and Waller, B. M. (2006). "Understanding chimpanzee facial expression: insights into the evolution of communication." *Social Cognitive and Affective Neuroscience*, 1(3), 221–8.

Patterson, F. G. (1978). "The gestures of a gorilla: Language acquisition in another pongid." *Brain and Language*, 5(1), 72–97.

Penn, D. C. (2011). "How folk psychology ruined comparative psychology: And how scrub jays can save it." In Randolf Menzel and Julia Fischer (eds.), *Animal Thinking: Contemporary Issues in Comparative Cognition*, 253–66. MIT Press.

Penn, D. C., and Povinelli, D. J. (2007a). "On the lack of evidence that non-human animals possess anything remotely resembling a 'theory of mind'." *Philosophical Transactions of the Royal Society B: Biological Sciences*, 362(1480), 731–44.

224 REFERENCES

Penn, D. C., and Povinelli, D. J. (2007b). "Causal cognition in human and nonhuman animals: A comparative, critical review." *Annual Review of Psychology*, 58(1), 97–118.

Pepperberg, I. M. (2009). *The Alex Studies: Cognitive and Communicative Abilities of Grey Parrots*. Harvard University Press.

Perner, J., and Lang, B. (1999). "Development of theory of mind and executive control." *Trends in Cognitive Sciences*, 3(9), 337–44.

Piccinini, G., and Craver, C. (2011). "Integrating psychology and neuroscience: Functional analyses as mechanism sketches." *Synthese*, 183(3), 283–311.

Ponte, G., Chiandetti, C., Edelman, D. B., Imperadore, P., Pieroni, E. M., and Fiorito, G. (2022). "Cephalopod behavior: from neural plasticity to consciousness." *Frontiers in Systems Neuroscience*, 15, 787139.

Povinelli, D. J. (1994). "Comparative studies of animal mental state attribution: A reply to Heyes." *Animal Behaviour*, 48, 239–41.

Povinelli, D. J. (2020). "Can comparative psychology crack its toughest nut?" *Animal Behavior and Cognition*, 7(4), 589–652.

Povinelli, D. J., and Eddy, T. J. (1996). *What Young Chimpanzees Know about Seeing*. University of Chicago Press.

Povinelli, D. J., Nelson, K. E., and Boysen, S. T. (1990). "Inferences about guessing and knowing by chimpanzees (Pan troglodytes)." *Journal of Comparative Psychology*, 104(3), 203.

Povinelli, D. J., and Vonk, J. (2003). "Chimpanzee minds: suspiciously human?" *Trends in Cognitive Sciences*, 7(4), 157–60.

Powell, R. (2007). "Is convergence more than an analogy? Homoplasy and its implications for macroevolutionary predictability." *Biology & Philosophy*, 22(4), 565–78.

Premack, D., and Woodruff, G. (1978). "Does the chimpanzee have a theory of mind?" *Behavioral and Brain Sciences*, 1(4), 515–26.

Putnam, H. (1975). "Philosophy and our mental life." In H. Putnam (ed.), *Mind, Language and Reality: Philosophical Papers* (Vol. 2, 291–303). Cambridge University Press.

Quine, W. V. O. (1948). "On what there is." *Review of Metaphysics*, 2(1), 21–38.

Quine, W. V. O. (1951). "Two dogmas of empiricism." *Philosophical Review*, 60, 20–43

Ramsay, M. S., and Teichroeb, J. A. (2019). "Anecdotes in primatology: Temporal trends, anthropocentrism, and hierarchies of knowledge." *American Anthropologist*, 121(3), 680–93.

Remane, A. (1952). *Die Grundlagen des natürlichen Systems, der vergleichenden Anatomie und der Phylogenetik* (repr. 1971). Otto Koeltz.

Rescorla, R.A., and Wagner, A. R. (1972). "A theory of Pavlovian conditioning: Variations in the effectiveness of reinforcement and nonreinforcement." In A. H. Black and W. F. Prokasy (eds.), *Classical Conditioning II*, 64–99. Appleton-Century-Crofts.

Ribbands, C. (1949). "The foraging method of individual honey-bees." *Journal of Animal Ecology*, 18, 47–66.

Rivas, J., and Burghardt, G. M. (2002). "Crotalomorphism: A metaphor for understanding anthropomorphism by omission." In M. Bekoff, C. Allen, and G. M. Burkhardt (eds.), *The Cognitive Animal: Experimental and Theoretical Perspectives on Animal Cognition*, 9–17. MIT Press.

Romanes, G. J. (1883). *Animal Intelligence*. D. Appleton & Co.

Romero, F. (2019). "Philosophy of science and the replicability crisis." *Philosophy Compass*, 14(11), e12633.

REFERENCES 225

Rosati, A. G., Wobber, V., Hughes, K., and Santos, L. R. (2014). "Comparative developmental psychology: how is human cognitive development unique?" *Evolutionary Psychology*, 12(2), 147470491401200211.

Rose, H., and Rose, S. (2010). *Alas Poor Darwin: Arguments against Evolutionary Psychology*. Random House.

Rosenbaum, R. S., Köhler, S., Schacter, D. L., Moscovitch, M., Westmacott, R., Black, S. E., ... and Tulving, E. (2005). "The case of KC: contributions of a memory-impaired person to memory theory." *Neuropsychologia*, 43(7), 989–1021.

Rosenberg A (2006) *Darwinian Reductionism: or, How to Stop Worrying and Love Molecular Biology*. University of Chicago Press.

Rosenthal-von der Pütten, A. M., Krämer, N. C., Hoffmann, L., Sobieraj, S., and Eimler, S. C. (2013). "An experimental study on emotional reactions towards a robot." *International Journal of Social Robotics*, 5(1), 17–34.

Ross, D. M. (1971). "Protection of hermit crabs (Dardanus spp.) from octopus by commensal sea anemones (Calliactis spp.)." *Nature*, 230(5293), 401–2.

Rouder, J. N., Speckman, P. L., Sun, D., Morey, R. D., and Iverson, G. (2009). "Bayesian t tests for accepting and rejecting the null hypothesis." *Psychonomic Bulletin & Review*, 16(2), 225–37.

Ruijten, P. A., Haans, A., Ham, J., and Midden, C. J. (2019). "Perceived human-likeness of social robots: testing the Rasch model as a method for measuring anthropomorphism." *International Journal of Social Robotics*, 11(3), 477–94.

Rutiku, R., Aru, J., and Bachmann, T. (2016). "General markers of conscious visual perception and their timing." *Frontiers in Human Neuroscience*, 10, 23.

Samuels, R. (1998). "Evolutionary psychology and the massive modularity hypothesis." *British Journal for the Philosophy of Science*, 49, 575–602.

Schirmer, A., Seow, C. S., and Penney, T. B. (2013). "Humans process dog and human facial affect in similar ways." *PLOS ONE*, 8(9), e74591.

Schmuckler, M. A. (2001). "What is ecological validity? A dimensional analysis." *Infancy*, 2(4), 419–36.

Scotland, R. W. (2010). "Deep homology: a view from systematics." *Bioessays*, 32(5), 438–49.

Scott, R. B., Samaha, J., Chrisley, R., and Dienes, Z. (2018). "Prevailing theories of consciousness are challenged by novel cross-modal associations acquired between subliminal stimuli." *Cognition*, 175, 169–85.

Seeley, T. D. (2010). *Honeybee Democracy*. Princeton University Press.

Seeley, T. D. (2014). *Honeybee Ecology: a Study of Adaptation in Social Life* (Vol. 36). Princeton University Press.

Seeley, T. D., Camazine, S., and Sneyd, J. (1991). "Collective decision-making in honeybees: how colonies choose among nectar sources." *Behavioral Ecology and Sociobiology*, 28(4), 277–90.

Seth, A. K., and Bayne, T. (2022). "Theories of consciousness." *Nature Reviews Neuroscience*, 23(7), 439–52.

Shaki, S., and Fischer, M. H. (2020). "Nothing to dance about: Unclear evidence for symbolic representations and numerical competence in honeybees. A comment on: Symbolic representation of numerosity by honeybees (Apis mellifera): Matching characters to small quantities." *Proceedings of the Royal Society B*, 287(1925), 20192840.

Shallice, T. (2015). "Cognitive neuropsychology and its vicissitudes: The fate of Caramazza's axioms." *Cognitive Neuropsychology*, 32(7–8), 385–411.

Shallice, T. (2019). "The single case study of memory." In S. E. MacPherson and S. Della Sala (eds.), *Cases of Amnesia*, 1–15. Routledge.

226 REFERENCES

Shaman, N. J., Saide, A. R., and Richert, R. A. (2018). "Dimensional structure of and variation in anthropomorphic concepts of God." *Frontiers in Psychology*, 9, 1425.

Sheridan, T. B. (2020). "A review of recent research in social robotics." *Current Opinion in Psychology*, 36, 7–12.

Shettleworth, S. (2010). *Cognition, Evolution, and Behavior*. 2nd edn. Oxford University Press.

Shettleworth, S. (2012). *Fundamentals of Comparative Cognition*. Oxford University Press.

Shevlin, H. (2021). "Non-human consciousness and the specificity problem: A modest theoretical proposal." *Mind & Language*, 36(2), 297–314.

Shubin, N., Tabin, C., and Carroll, S. (2009). "Deep homology and the origins of evolutionary novelty." *Nature*, 457(7231), 818–23.

Siegel, A. (2004) *Neurobiology of Aggression and Rage*. CRC Press.

Siegel, A., and Victoroff, J. (2009) "Understanding human aggression: new insights from neuroscience." *International Journal of Law and Psychiatry*, 32(4):209–15.

Silverman, P. S. (1997). "A pragmatic approach to the inference of animal minds." In R. W. Mitchell, N. S. Thompson, and H. L. Miles (eds.), *Anthropomorphism, Anecdotes, and Animals*, 170–85. State University of New York Press.

Smith, J. A., and Boyd, K. M. (1991). *Lives in the Balance: the Ethics of Using Animals in Biomedical Research: the Report of a Working Party of the Institute of Medical Ethics*. Oxford University Press.

Smith, J. D., Couchman, J. J., and Beran, M. J. (2014). "Animal metacognition: a tale of two comparative psychologies." *Journal of Comparative Psychology*, 128(2), 115.

Smith, L. B. (2000). "Avoiding associations when it's behaviorism you really hate." In R. Golinkoff and K. Hirsh-Pasek (eds.), *Breaking the Word Learning Barrier*, 169–74. Oxford University Press.

Smith, P. L., and Little, D. R. (2018). "Small is beautiful: In defense of the small-N design." *Psychonomic Bulletin & Review*, 25(6), 2083–101.

Smith, S. E. (2020). "Is evolutionary psychology possible?" *Biological Theory*, 15(1), 39–49.

Sneddon, L. U., Elwood, R. W., Adamo, S. A., and Leach, M. C. (2014). "Defining and assessing animal pain." *Animal Behaviour*, 97, 201–12.

Sober, E. (1990). "Explanation in biology: Let's razor Ockham's Razor." *Royal Institute of Philosophy Supplements*, 27, 73–93.

Sober, E. (1991). *Reconstructing the Past: Parsimony, Evolution, and Inference*. MIT Press.

Sober, E. (1998). "Morgan's canon." In Denise Cummins and Colin Allen (eds.), *The Evolution of Mind*, 224–42. Oxford University Press.

Sober, E. (2005). "Comparative psychology meets evolutionary biology." In L. Daston and G. Mitman (eds.), *Thinking with Animals: New Perspectives on Anthropomorphism*, 85–99. Columbia University Press.

Sober, E. (2012). "Anthropomorphism, parsimony, and common ancestry." *Mind & Language*, 27(3), 229–38.

Sober, E. (2015). *Ockham's Razors: A User's Manual*. Cambridge University Press.

Soll, J. B., Milkman, K. L., and Payne, J. W. (2015). "A user's guide to debiasing." *The Wiley Blackwell Handbook of Judgment and Decision Making*, Part II, 924–51.

Solvi, C., Baciadonna, L., and Chittka, L. (2016). "Unexpected rewards induce dopamine-dependent positive emotion–like state changes in bumblebees." *Science*, 353(6307), 1529–31.

Sommer, R. J. (2008). "Homology and the hierarchy of biological systems." *BioEssays*, 30(7), 653–8.

Soto, F. A., Vogel, E. H., Uribe-Bahamonde, Y. E., and Perez, O. D. (2023). "Why is the Rescorla-Wagner model so influential?" *Neurobiology of Learning and Memory*, 204, 107794.

Spatola, N., Belletier, C., Chausse, P., Augustinova, M., Normand, A., Barra, V., . . . and Huguet, P. (2019). "Improved cognitive control in presence of anthropomorphized robots." *International Journal of Social Robotics*, 11(3), 463–76.

Spaulding, S. (2013). "Mirror neurons and social cognition." *Mind & Language*, 28(2), 233–57.

Stanovich, K. E., and West, R. F. (2000). "Individual differences in reasoning: Implications for the rationality debate?" *Behavioral and Brain Sciences*, 23(5), 645–65.

Stanton, L., Davis, E., Johnson, S., Gilbert, A., and Benson-Amram, S. (2017). "Adaptation of the Aesop's Fable paradigm for use with raccoons (Procyon lotor): considerations for future application in non-avian and non-primate species." *Animal Cognition*, 20(6), 1147–52.

Starzak, T. B., and Gray, R. D. (2021). "Towards ending the animal cognition war: a three-dimensional model of causal cognition." *Biology & Philosophy*, 36(2), 1–24.

Staudte, M., and Crocker, M. W. (2009). "Visual attention in spoken human-robot interaction." In *4th ACM/IEEE International Conference on Human-Robot Interaction (HRI)*, 77–84. IEEE.

Stevens, J. R. (2017). "Replicability and reproducibility in comparative psychology." *Frontiers in Psychology*, 8, 862.

Stout, S. C., and Miller, R. R. (2007). "Sometimes-competing retrieval (SOCR): a formalization of the comparator hypothesis." *Psychological Review*, 114(3), 759.

Striedter, G. F., and Northcutt, R. G. (1991). "Biological hierarchies and the concept of homology." *Brain, Behavior and Evolution*, 38(4–5), 177–89.

Strube-Bloss, M. F., Nawrot, M. P., and Menzel, R. (2011). "Mushroom body output neurons encode odor–reward associations." *Journal of Neuroscience*, 31(8), 3129–40.

Sullivan, J. A. (2022). "Novel tool development and the dynamics of control: The rodent touchscreen operant chamber as a case study." *Philosophy of Science*, 89(5), 1203–12.

Tal, E. (2017). "Calibration: Modelling the measurement process." *Studies in History and Philosophy of Science Part A*, 65, 33–45.

Tautz, J., Zhang, S., Spaethe, J., Brockmann, A., Si, A., Srinivasan, M., and Collett, T. (2004). "Honeybee odometry: performance in varying natural terrain." *PLOS Biology*, 2(7), e211.

Taylor, A. H., Bastos, A. P., Brown, R. L., and Allen, C. (2022). "The signature-testing approach to mapping biological and artificial intelligences." *Trends in Cognitive Sciences*, 26(9), 738–50.

Thorndike, E. L. (1911). *Animal Intelligence: Experimental Studies*. Macmillan.

Timberlake, W. (2002). "Niche-related learning in laboratory paradigms: the case of maze behavior in Norway rats." *Behavioural Brain Research*, 134(1–2), 355–74.

Tomasello, M., and Call, J. (2006). "Do chimpanzees know what others see—or only what they are looking at?" In S. Hurley and M. Nudds (eds.), *Rational Animals?* 371–84. Oxford University Press.

Tomasello, M., and Moll, H. (2013). "Why don't apes understand false beliefs?" In M. Banaji and S. Gelman (eds.), *Navigating the Social World: What Infants, Children, and Other Species Can Teach Us*, 81–8. Oxford University Press.

Travers, E., Frith, C. D., and Shea, N. (2018). "Learning rapidly about the relevance of visual cues requires conscious awareness." *Quarterly Journal of Experimental Psychology*, 71(8), 1698–713.

228 REFERENCES

Van Doorn, J., Mende, M., Noble, S. M., Hulland, J., Ostrom, A. L., Grewal, D., and Petersen, J. A. (2017). "Domo arigato Mr. Roboto: Emergence of automated social presence in organizational frontlines and customers' service experiences." *Journal of Service Research,* 20(1), 43–58.

Van Gulick, Robert. (2021). "Consciousness." In *The Stanford Encyclopedia of Philosophy* (Winter 2021 Edition), ed. Edward N. Zalta, https://plato.stanford.edu/archives/win2 021/entries/consciousness/.

Van Hamme, L. L., and Wasserman, E. A. (1994). "Cue competition in causality judgments: the role of nonpresentation of compound stimulus elements." *Learning and Motivation,* 25, 127–51.

Visalberghi, E. (2018). "Trap-tube problem." In J. Vonk and T. Shackelford (eds.), *Encyclopedia of Animal Cognition and Behavior.* Springer, https://doi.org/10.1007/978-3-319-47829-6_1484-1.

Visscher, P. K., and Seeley, T. D. (1982). "Foraging strategy of honeybee colonies in a temperate deciduous forest." *Ecology,* 63(6), 1790–1801.

Von Frisch, K. (1974). "Decoding the language of the bee." *Science,* 185(4152), 663–8.

Vonk, J., and Beran, M. J. (2022). "Outgoing editors' reflections." *Animal Behavior and Cognition,* 9(3), 257–60.

Vonk, J., and Krause, M. A. (2018). "Editorial: Announcing preregistered reports." *Animal Behavior and Cognition,* 5(2).

Vonk, J., and Shackelford, T.K. (2013). "An introduction to comparative evolutionary psychology." *Evolutionary Psychology,* 11, 459–69.

Waddington, K. D., and Holden, L. R. (1979). "Optimal foraging: on flower selection by bees." *American Naturalist,* 114(2), 179–96.

Wagenmakers, E. J., Lee, M., Lodewyckx, T., and Iverson, G. J. (2008). "Bayesian versus frequentist inference." In H. Hoijtink, I. Klugkist, and P. Boelen (eds.), *Bayesian Evaluation of Informative Hypotheses,* 181–207. Springer.

Wagenmakers, E. J., Wetzels, R., Borsboom, D., and Van Der Maas, H. L. (2011). "Why psychologists must change the way they analyze their data: the case of psi: comment on Bem." *Journal of Personality and Social Psychology,* 100(3):426–32.

Wagner, G. P. (2014). *Homology, Genes, and Evolutionary Innovation.* Princeton University Press.

Wagner, G. P. (2016). "What is 'homology thinking' and what is it for?" *Journal of Experimental Zoology Part B: Molecular and Developmental Evolution,* 326(1), 3–8.

Wall, R. J., and Shani, M. (2008). "Are animal models as good as we think?" *Theriogenology,* 69(1), 2–9.

Waller, B. M., and Dunbar, R. I. (2005). "Differential behavioural effects of silent bared teeth display and relaxed open mouth display in chimpanzees (Pan troglodytes)." *Ethology,* 111(2), 129–42.

Waytz, A., Cacioppo, J., and Epley, N. (2010). "Who sees human? The stability and importance of individual differences in anthropomorphism." *Perspectives on Psychological Science,* 5(3), 219–32.

Waytz, A., Epley, N., and Cacioppo, J. T. (2010). "Social cognition unbound: Insights into anthropomorphism and dehumanization." *Current Directions in Psychological Science,* 19(1), 58–62.

Waytz, A., Heafner, J., and Epley, N. (2014). "The mind in the machine: Anthropomorphism increases trust in an autonomous vehicle." *Journal of Experimental Social Psychology,* 52, 113–17.

Waytz, A., Morewedge, C. K., Epley, N., Monteleone, G., Gao, J. H., and Cacioppo, J. T. (2010). "Making sense by making sentient: effectance motivation increases anthropomorphism." *Journal of Personality and Social Psychology*, 99(3), 410.

Weisberg, M. (2013). *Simulation and Similarity: Using Models to Understand the World*. Oxford University Press.

Wellman, H. M., Cross, D., and Watson, J. (2001). "Meta-analysis of theory-of-mind development: The truth about false belief." *Child Development*, 72(3), 655–84.

Wenzel, J. W. (1992). "Behavioral homology and phylogeny." *Annual Review of Ecology and Systematics*, 361–81.

Wiegman, I. (2016). "Angry rats and scaredy cats: lessons from competing cognitive homologies." *Biological Theory*, 11(4), 224–40.

Woodruff, G., and Premack, D. (1979). "Intentional communication in the chimpanzee: The development of deception." *Cognition*, 7(4), 333–62.

Worden, B. D., and Papaj, D. R. (2005). "Flower choice copying in bumblebees." *Biology Letters*, 1(4), 504–7.

Wray, M. K., Klein, B. A., Mattila, H. R., and Seeley, T. D. (2008). "Honeybees do not reject dances for 'implausible' locations: reconsidering the evidence for cognitive maps in insects." *Animal Behaviour,* 76(2), 261–9.

Wright, J. T., and Gaudi, B. S. (2013). "Exoplanet detection methods." In L. M. French and P. Kalas (eds.), *Planets, Stars and Stellar Systems, Vol. 3: Solar and Stellar Planetary Systems*, 489–540. Springer.

Wynne, C. D. (2004). "The perils of anthropomorphism." *Nature*, 428(6983), 606.

Wynne, C. D. (2007). "What are animals? Why anthropomorphism is still not a scientific approach to behavior." *Comparative Cognition & Behavior Reviews*, 2, 125–35.

Yarkoni, T. (2022). "The generalizability crisis." *Behavioral and Brain Sciences*, 45.

Ylikoski, P., and Kuorikoski, J. (2010). "Dissecting explanatory power." *Philosophical Studies*, 148(2), 201–19.

Złotowski, J., Sumioka, H., Eyssel, F., Nishio, S., Bartneck, C., and Ishiguro, H. (2018). "Model of dual anthropomorphism: the relationship between the media equation effect and implicit anthropomorphism." *International Journal of Social Robotics*, 10, 701–14.

Index

For the benefit of digital users, indexed terms that span two pages (e.g., 52–53) may, on occasion, appear on only one of those pages.

Tables and figures are indicated by an italic *t* and *f* following the page numbers. Page numbers followed by n denote notes.

acetic acid 184–85
adaptationism 115
affective priming 64–66
affect programs 103–5
aggression 104–7
agricultural model 136–38
Alem, S. 144–45
Allen, C. 71–72, 73, 75–76, 77, 135n.6, 170, 207
'alternative' hypothesis 26–27
Alupay, J. S. 184
amygdala 104–5, 106–7
analogous traits 98–99
anchoring experiments 141–47
Anderson, M. L. 182
Andrews, K. 35, 37, 161–62
anecdotal evidence 150–52, 161–62, 167–69
anger 102, 103, 104–5
  with fear 106–7
  homology of 105–7
Animal Behavior and Cognition (journal) 154–55, 158
Animal Cognition (journal) 154–55
animals
  causal cognition 38
  cognition models 18
  sentience 38
  welfare 75–76, 174, 176
anthropocentric intuition 170
anthropodenial 53, 66–67
anthropomorphism
  as an error *vs.* appropriate attribution 49–50

automatic responses to cues 62
bias 51–52, 55–57, 60–64
Darwin's influence 48
errors of commission and omission 52–54
evaluation of models 56
God concept and 61–62
implicit anthropomorphism 54–56, 64–66
interpretations 167–71
introduction 46–48
methods for studying 60–64
Morgan's Canon 52, 55–56
philosophical debates and practical implications of 66–69
as a process 50–51
psychological study of 57–58
three-factor theory 58–60
varied perspectives on 46–48
anxiety 54
apes 100–1, 119, 132, 144
Apperly, I. A. 36, 37, 40
Arico, A. 58–59
arthropods 174
artificiality to experiments
  in background of the animals 132
  unusual life histories and training and 132–33
artificial stimuli 124–26
associationism 35, 89–90, 91
association *vs.* cognition 72–73
associative/cognitive dichotomy 89–91, 135n.7
associative learning 83, 144–45, 164, 176–77, 189

232 INDEX

associative models 72–73, 188–89
  abstract description view 83–84, 86, 89–90, 91
  associative processing 81–82
  challenges in 82–83
  vs cognitive models 76–77, 80
  development of 78–79
  interpretation of 83–86
  mechanistic explanation and 86–93
  'thinking' models 83
associative processing 81–82, 83–84, 85
  defined 92
  mechanistic explanation and 87
'asymmetric dependency' of mind-reading abilities 23–24
automatic vs controlled processing 40
auxiliary hypotheses 15–16
Ayumu (chimpanzee) 153–54, 156–57, 169, 170

Baars, B. J. 179–80
backward blocking 79
Barrett, J. L. 61–62, 103
Barron, A. B. 180–81
basic emotions 108
bats 100–1
Bausman, W. 29–30
Bayesianism 93, 203, 204–5 see also likelihoodism
  Bayesian brain/predictive processing 72
  Bayesian inference 203, 205, 208
  Bayesian statistics 201, 204–5
Bayne, T. 190
bees 70, 145–46
  arithmetic study 134–35, 146–47
  biases in behavior 185
  cognitive flexibility in 133n.5, 134–35
begging behaviors in chimpanzees 19–20
behavioral genetics 120
behaviorism 2, 16, 55, 75–76, 112n.15
  Dennett and 112n.15
behaviorists 78, 81–82, 131
behavior reading 22, 23–24, 25, 28, 29, 34, 36–37, 39–42, 41f, 52, 72–73, 111–12
Bekoff, M. 61–62
Beran, M. J. 76
Birch, J. 38, 176, 177–78, 182, 186, 189
Blaisdell, A. P. 80
Blanchard, D. C. 105

Blanchard, R. J. 105
Block, Ned 173n.1
body orientation 19–20, 23–24, 38–39
bonobos 21, 100–1
boundary problem for cognitive homologies 97–98, 102, 105–8, 112, 113, 115–16, 118, 120
Boyle, A. 159–60
Boysen, S. T. 19
Brownell, Abe 156f
Brunswik, Egon 138–40, 159
Bublitz, A. 183
Buckner, Cameron 39, 41–42, 76, 89–92
Bullock, T. H. 183–84
bumblebees 134–35, 143, 144–45
Burghardt, G. M. 49–52, 55
Butterfill, S. A. 36, 37, 40

Cacioppo, J. T. 58, 59, 60n.12
calibration 35, 37
Call, J. 24
Camazine, S. 142
Campbell, D. T. 165
Caramazza, A. 163
care 104–5
Carroll, N. C. 64–65
Carruthers, P. 181n.7
case-based reasoning 164–65
case studies 164–66
cats 103–7
causal cognition 74, 75, 144
  association and 83–84
  modeling 75
causal graphical models 75, 81f
causal learning 75
causal reasoning 72, 74–75, 83–84, 91–92
  ability 84
  contrasting with association 74–75
  dimensional model of 77
  in rats 72–73
causal structure and inference 79–80
causal sufficiency 74–75, 78–79
central association unit 187–88
cephalopods
  cognitive abilities of 55
  consciousness 175–77, 195
Chalmers, D. J. 173–74
cheater detection module 114–15, 116
Cheke, L. G. 144–45

chemical cues 182
Cheney, D. L. 145–46
chimpanzees 110, 156–57, 169
    forelimb 100–1
    mind-reading behavior 110–13
    performance on a short-term memory
        task 153–54
Chomsky, Noam 71
cladistic parsimony 116–17
cladistic spread of the trait 100–1
classical conditioning 36, 131, 144
class-specificity constraint 107
Clayton, N. S. 38, 144–45, 159–60, 204–5
Coane, Jen 64–65
cognitive arousal 63
cognitive biases 47, 48, 55–56, 67–69
cognitive capacities 30, 38, 97, 101–2, 118,
        119, 120, 140, 162, 163
cognitive cartography 43–44
cognitive emotions 103
cognitive ethologists 133, 134–35
cognitive ethology 123, 133, 160–61
    incident reports in 123, 133, 160–62
cognitive flexibility 133n.5, 134–35
cognitive neuropsychology, lesion studies
        in 162–63
cognitive processing 92
    cognitive models 71–73, 74, 76–78, 83–
        85, 91, 204
communication in insects 126
comparative cognition 77
comparative psychologists 133, 134
comparative psychology 16, 30–31
    comparative inferences and 97
    models in 38
    negative results in 27–28
comparative psychology vs cognitive
        ethology 123
competitive feeding experiment 23,
        111n.14
competitive feeding paradigm 20–21
complimentary behavior-reading
        hypothesis 23–24, 36–37
computational models 75–76
computational neuroscience 89–91
connectionism 72
connectionist models 91
consciousness in animals
    cephalopodas 175–77

clusters and clutters 189–91
global workspace theory 179
hard problem of 173–74
indicators and measures 185–86
methodological approach to
        study 174–75
midbrain theory 180, 187–88
multiple clusters 191–95
pain experience 184–85
perceptual consciousness 182–84
philosophical perspectives 173–74
scientific study of 174
subjectivity 180n.6
theory and consciousness 177–78
theory-heavy approaches 179–82
theory-light approaches 182
unlimited associative learning
        (UAL) 187–89
context
    of discovery 56n.9
    of justification 56n.9
contingent interaction with the
        environment 58–59
coordination problem of measurement 185
Cosmides, L. 114–15
Couchman, J. J. 76
crabs 174
critical anthropomorphism 49–50
Crocker, M. W. 63
Crook, R. J. 184–85
Cross, E. S. 62–63
cross-modal learning 182
cues 17–18, 21, 23–24, 38–39, 65–66,
        138–39, 182
cuttlefish 175, 180–81, 183–85

Dacey, M. 64–65, 76, 83–84, 89–91
Darwin 150–52
debiasing 66, 70
deep homologies 118–19
default model/hypothesis 26–29
defense system 105
demandingness problem 181
Dennett, D. C. 15, 112n.15
developmental psychology 21–22, 120,
        129, 144
de Waal, F. B. 53–54, 110, 119
dimensional modeling 77, 204
direct reports method 60

234 INDEX

disgust 102
dissociation logic 169–70
distance perception 139
distribution question 173
dolphins 100–1
Dretskian view of representational
content 41–42
dual systems hypothesis 40
Dwyer, D. M. 80
dynamical systems theory 72

ecological validity
anchoring experiments 141–47
artificiality, background of the animals
and 132–33
artificiality of experiments 122, 127–30
comparative psychology vs cognitive
ethology 123
compiling experiments across
situations 138–40
defense of external invalidity 136–38
definition and importance 122
dimensions 130–32
disciplinary differences 123
honeybee cognition, experiments
with 124–27
lab vs field experiments 133–36
ecumenicism 43
Eddy, T. J. 19–20, 58–59
Edelman, D. B. 179–80
effectance motivation 58, 59
egalitarianism 68–69
electroencephalogram (EEG) 183–84
elicited agent knowledge 58–59
elimination of mental entities 75–76
eliminative materialism 75–76
eliminativism 92–93
emotional arousal 106–7
emotional response, physiological markers
of 62–63
emotions 98
affect programs 103–5
anger 64–65, 102, 103, 104–7
anxiety 54
basic 102–3, 108
constructivism 103
disgust 64–65
emotional arousal 106–7
fear 64–65

happiness 64–65
moods 17–18, 185, 192*f*, 193*f*
rage 104–7, 108
recognition of 64–65
sadness 64–65
empathetic response of participants 62–63
empathy-driven perspective-taking
approach 35
empirical cluster 191
empirical conception of parsimony
claims 32–34
empirical parsimony 32–34, 33*t*
empiricism 102n.7
environmental context 133–34
episodic memory 162
epistemic clusters 191–94
epistemic considerations 32–33
Epley, N. 58, 59, 60n.12, 62–63, 66
ethologists 133–35
ethology 123, 133, 134, 160–62
evidentialism 31–32
evidential value of lesion studies 162–63
evolutionary psychology
boundary problems 115–16
cheater detection module 114–15
comparative inferences 116–17
core claims 114–16
criticisms of 115
definition and scope 114
explanatory depth 86
explicit anthropomorphism 65–66
explicitness 38–39
explicit report methods 61
extended comparator model 79
external validity 145
eye tracking of participants 63

face recognition 40–41
facial expressions 23–24, 62–63, 64–
65, 103
facial features 58–59
facilitation hypothesis 178, 189
false belief in mind-reading 15, 21–22,
36, 41, 43
false belief task 144
Farrar, B. 159–60, 204–5
fear 102, 104–5, 106–7
Ferrari, P. F. 119
field research 133–34

INDEX

Fischer, M. H. 124–25
fish 98–99, 174
Fitzpatrick, S. 31–32
folk-psychological terminology 75–76
food stimulus 78
foraging behaviors 124–27
Forrester, John 164–65
forward blocking 78–79
Frith, C. D. 183

Gallup, G. G. 58–59
gastropod mollusks 185
gaze and behavior 21
gene networks 120
generalizability 136–38, 145, 170
   replicability and 150, 156–58, 159–60,
     164–65, 166, 168–71
generalizations 168–70
   of experimental results 74, 140
   of theories 136–38
geometric orientation and mental
   state 21
gibbons 100–1
Ginsburg, S. 178, 187–89
global workspace theory 179, 193–94
Godfrey-Smith, P. 180–81
gorillas 100–1, 170–71
Gould, C. G. 126–27
Gould, J. L. 126–27
Gray, K. 38–39, 77, 93
gray parrot 170–71
Griffiths, P. E. 103

Hadjisolomou, S. P. 184
Halina, Marta 29–30, 159, 177–78,
   186, 189–90
hamate bone 99n.5
happiness 102
Hare, B. 20–21, 28, 111n.14
Harrison, D. 177–78, 186, 189–90
Heafner, J. 62–63
Heider, F. 61–62
hemocyanin 110n.12
Heyes, C. 21–22, 22n.4, 28–29, 36, 37, 40,
   42–43, 76
hippocampus 88f
Hoffmann, L. 144–45
homeostatic property cluster 190
homologous traits 96–97, 98–99, 100–2

homology
   of anger 105–7
   basics 96–102
   boundary problem 105–8
   challenge of 97–98, 103n.8, 103
   cladistic parsimony 109–13
   cognitive/functional 97, 98–100
   comparative inferences 97–98
   continuity of intermediates 99–100,
     107, 119
   cross-disciplinary integration 96–97
   evolutionary psychology and its
     limits 114–17
   gene regulatory network 100n.6
   multi-level understanding of 118–19
   neuroscience of emotions 102–5
   position 100
   special character 100
   structural 96–97, 98–100
homology-based comparative claims 117f
honey 80
honeybees
   foraging experiment 142–44
   numerical cognition in 124–27
   waggle dance 142–44
'horizontal' homology 115
Horowitz, A. C. 61–62
horseshoe crabs 110n.12
hoses 100–1
Howard, S. R. 124, 128–29, 134–35
human folk psychology 57–58
humans
   arms and bird wings,
     commonalities 96–97
   basic emotions in 102–3
   bird forelimbs and 99–100
   uniqueness 2
Hume, D. 38, 47, 81–82
hypothalamus 104–5, 106–7

implicit anthropomorphism 47–48, 50–
   52, 65–66
   awareness and study of its effects 69
   intuitive folk psychology and 74
implicit association test 63
implicit folk psychology 50n.3, 51–52, 69
implicit manipulation 63
indirect reports method 60
individual associative model 84

236 INDEX

individual difference in anthropomorphism questionnaire (IDAQ) 60–61, 63, 65, 66
'individual spread' of a trait 100–1
inference and causal structure 79–80
inference to best explanation 31f, 41–42, 56–57, 197–98, 203, 205, 208
inferential gap 130
informed experience 54n.8
innateness 190
Inoue, S. 153–54, 156–57, 169, 170
input–output mappings 140
insects 174, 180–81
integration
  cognitive science and 95–96
  cross-disciplinary 96–97
  multimodal 143
  themes and challenges in 118–20
intentional (critical) anthropomorphism 51
internal validity 137–38, 145
interpretation of evidence 3
intuitive folk psychology 55, 68, 74
Irvine, E. 188

Jablonka, E. 178, 187–89
Journal of Comparative Psychology 154–55
Journal of Experimental Psychology: Animal Learning and Cognition 154–55

Kanet, S. 43
Kano, F. 22n.4
Keil, F. C. 61–62
Kennedy, J. S. 50–52
Klein, B. A. 177–78, 180–81, 186, 189–90
Koltermann, R. 143
Krachun, C. 43
Krupenye, C. 21, 23, 35, 36
Kuhn, T. 1, 77n.2

laboratory psychology 133–34
lesion studies in cognitive neuropsychology 162–63
Levine, J. 173–74
Li, Z. 63
lidocaine 184–85
likelihoodism 203, 204–5
limited associative learning (LAL) 187
Little, D. R. 163–64

lobsters 174
loneliness 59
Loukola, O. J. 134–35
Love, A. C. 100
Lugrin, B. 62–63
Lurz, R. W. 25, 43
lust 104–5

MaBouDi, H. 124–25, 128–29
Mackintosh, J. 183
Mackintosh, N. J. 183
Mameli, M. 190
ManyBirds Project 158–59
ManyDogs Project 158–59
ManyPrimates Project 158–59
Marchesi, S. 63
Marr, D. 91
massive modularity 114, 115
mathematical cognitive models 84–85
mathematical modeling 75–76
Mather, J. A. 177, 179–80, 182
Matsuzawa, T. 153–54, 156–57, 169, 170
Maxwell, James Clerk 71
maze learning 124–26, 131, 146–47
McLean, R. 184
measurement theory 185
mechanistic explanation 72, 86–93
memory 153, 162, 169
Menne, I. M. 62–63
mental entities, elimination of 75–76
mental geography 38
Merker, B. 180–81
mice 132
midbrain view and reafference 180
Mikhalevich, I. 76
Milkman, K. L. 70
mind-perception in human beings *see* anthropomorphism
mindreading 16
  attributions 35
  behavior-reading 22, 23–24, 25, 28, 29, 34, 36–37, 39–43, 41f, 52, 72–73, 111–12
  competitive feeding paradigm 20–21
  debate, parsimony claims in 33–34
  false belief task 144
  knower/ guesser paradigm 22–23
  logical problem of 15–26
  minimal 36, 37
  submentalizing 36, 37, 42–43

INDEX 237

mind-reading capacity models 37, 38–39
  psychosemantics 39–40
  systems-level claims 40
mirror neurons 119
misinterpretation of animal behaviors 54
Mitchell, R. W. 161–62
model fitting 164
modeling cognitive processes
  associative models, reinterpreting 72–73
  causal reasoning 72–73, 77–80
  comparative psychology
    challenges 71–72
  interpreting associative models 81–86
  mechanistic explanation and the aims of
    modeling 86–93
  models in comparative psychology,
    themes and critiques of 73–77
  scientific advances and models 71
modesty 10–11, 197–98, 202
  about experimental results 29–32, 208
  about parsimony claims 32–34, 208
modularity 101–2, 107, 114, 115, 116
mollusks 175
monkeys 119, 151
moods 17–18, 185, 192*f*
Mook, D. G. 136–38
Morewedge, C. K. 58–59
Morgan, C. Lloyd 48, 151
Morgan, M. S. 164–66
Morgan's Canon
  anthropomorphism and 52–57, 66
  modeling and 76, 78, 82
  parsimony and 27, 29–30, 32–33, 33*t*
motivational trade-offs 184
motor learning 144–45
multilevel mechanistic explanation 72
multimodal integration 143
multisensory associative learning 188
multisensory learning 182

Nagel, T. 173–74
natural bias 51
Nature and Current Biology
  (journal) 155–56
negative results 27–28, 128, 129, 205
Nelson, N. C. 19, 123
Nemati, F. 56n.9
neobehaviorism 50–51
neural associationism 91
neural circuit models 91

neural correlates of consciousness 183–84
New York Declaration on Animal
  Consciousness 174
nociception information 184
nociceptive sensitization 185
non-anthropomorphic questions 60–61
noncompound stimuli 187
nonlinear movement trajectories,
  displaying 58–59
null hypothesis statistical testing
  (NHST) 26–28, 204, 208
  null hypothesis 29
numerical cognition in honeybees 124–27

octopuses
  cognitive abilities of 55, 70
  consciousness in 175–77, 179–81,
    182, 195
offense system 105
operant conditioning 144, 146
opposable thumbs 101–2
optimistic agnosticism 43
orangutans 21, 100–1
overestimation of animal intelligence 53–
  54, 61

Palmer, T. D. 182
panic 104–5
Panksepp, J. 103–5, 106–7, 108
Papineau, D. 76
parsimony
  cladistic 109–13, 116–17
  claim and underdetermination
    problem 26–29
  empirical 32–34, 33*t*
  formal 33*f*
  of mentalistic explanations 24–25
  modesty about parsimony claims 32–
    34, 208
  Morgan's Canon as 27, 29–30, 32–
    33, 33*t*
  of performance/ mechanism
    mapping 33*t*
  process count 33*f*
  of processing demands 33*t*
  simplicity 26–27
partial models 207, 208
Pavlov, Ivan 144
Payne, J. W. 70
Penn, D. C. 24, 28, 40, 74, 76

238 INDEX

perceived intelligence 65–66
perception 75–76, 139
perceptual access 19
perceptual capacities 55
perceptual states and goals 35
performance-mechanism mapping,
    parsimony of 32–33
periaqueductal gray 104–5
phenomenal consciousness 173n.1
physiological markers of emotional
    response 62–63
pigeons 146, 164
play 104–5
pooling
    multiple species 158–59
    of results 140
Popperian objection 27n.8, 27–28
posture 23–24
potential behaviors and actual behaviors,
    determining 137–38
Povinelli, D. J. 19–20, 23–24, 28, 30–31,
    35–36, 40, 58–59, 76
practical problems in
    underdetermination 17–18
pragmatic considerations 32
predation system 104–5
Premack, D. 15, 35
Preston, J. 58–59
primate grin 54, 61
priming 64–66
probabilistic causal reasoning 112n.16
process count parsimony 32–33, 33t
proof-of-concept 130
pseudo-nulls 29–30
psychiatric disorders 120
psychological autonomy thesis 91
psychological capacities 38, 101–2
psychological processing 92
psychosemantics 39–40
Putnam, H. 91
Pütten, Rosenthal-von der 62–63

Quine–Duhem thesis 15–16

raccoons 171–72
rage 104–5, 108
Ramsay, M. S. 161
Ramsey, A. K. 182
rapid reversal learning 183

rats 72, 103–4, 132
    causal reasoning in 72–73
    offense system in 105–7
    operant conditioning experiments
        with 146
reafference 180–81
reasoning 55, 69, 74–76
reductionism 32, 91
reductive/ neural associationism 89–90
registered reports 158, 160, 167, 171–72, 206
reinforcement 2, 132–33
religious intuitions 47–48
Remane, A. 99–100
replicability and generalizability 150, 156–
    58, 159–60, 164–65, 166, 168–71
replication crisis 157–60
representative design 138n.8
reptiles 174
Rescorla–Wagner (RW) model 78–79
research setting/ context, nature of 131
reversal learning 185–86
Rivas, J. 50–52, 55
Romanes, George 151–52
Ross, D. M. 184
Ruijten, P. A. 60–61

sadness 102
sample sizes, experimental
    animal anecdotes 150–52
    case studies in medicine and social
        science 164–66
    incident reports in cognitive
        ethology 161–62
    interpreting 167–71
    issues with 149
    lesion studies in cognitive
        neuropsychology 162–63
    replicability and generalizability 150
    replication crisis 157–60
    sample sizes in animal labs 153–57
    small-N design in vision science 163–64
Schmuckler, M. A. 130, 132
Schnell, A. 38
scrub jays 144–45
seeking 104–5
Seeley, T. D. 142
semantic memory 162
Seth, A. K. 179–80
Seyfarth, R. M. 145–46

Shaki, S. 124–25
shape recognition 40–41
Shea, N. 183, 190
Shevlin, H. 178
Simmel, M. 61–62
Skinner, B. F. 144
Sloan Committee meeting and report 95
"Sloan hexagram" of interdisciplinary
connections 96f
Smith, J. D. 76, 115, 163–64
Sneyd, J. 142
Sober, E. 32, 53–54, 110, 111, 112, 113,
116, 117
social agency 64
social agent, presence of 21
social biases 55, 68–69, 70
social cognition 36, 59, 112–13
social connections 59
social context 133–34
social coordination 116
sociality motivation 58, 59
socialization of animals and artificiality to
experiments 132–33
'social reasoning' system 40
social robotics 47–48, 64
social rule-breaking 114–15
Soll, J. B. 70
Spatola, N. 63
specificity trade-off in homology 97–
98, 117–18
spread of a trait 100–1
Stanton, L. 171–72
Starns, J. 80
Starzak, T. B. 38–39, 77, 93
statistical hypothesis 30–31
statistical inference 28, 31
statistical methods 156–57, 204–6
statistical nulls 29–30
statistical variance in behavioral
experiments 17–18
statistical vs substantive hypotheses 29–32
Staudte, M. 63
stereotype 55, 70
Stevens, J. R. 158–60
stimuli and materials, nature of 131
stimulus-response conditioning 2
strong vertical homologies 115
Stroop tests (of attentional control) 63
structural homology 96–97, 98–100

submentalizing 36, 37, 42–43
substantive hypothesis 30–31
success testing 11, 12–13, 17, 27, 42, 77n.2,
198, 199–200, 201, 206, 207, 208
signature testing 12–13, 206
surprise 102
Syntactic Structures (Chomsky) 71

target capacity 128–29
task, behavior, or response, nature of 132
Tautz, J. 126–27
Taylor, A. H. 42, 206
Teichroeb, J. A. 161
teleology 41
temporal discounting functions 164
testings prediction 137
theoretical problems in
underdetermination 18
theory and consciousness 177–78
theory-heavy approaches 185–86, 188,
191, 193
multiple cluster and 192–95
theory-light approaches 185–86, 191–92
multiple cluster and 192–95
Thorndike, Edward 48, 151
three-factor theory 66
Timberlake, W. 146
Tomasello, M. 24
Tooby, J. 114–15
training of animals and artificiality to
experiments 132–33
transfer tests 38–39
trap tube task 144
Travers, E. 183
trophy hunting 77, 77n.2, 82–83,
135n.6, 170
two-stage evidentialism 31f

umbrella hypotheses 22
unconscious processing 68n.14, 68–69
underdetermination 75
challenge: inferring into the black
box 17–19
chimpanzee mind-reading debates 19–22
parsimony claim and 26–29
practical problems in 17–18
Premack and Woodruff
experiment 15–16
theoretical problems in 18

240 INDEX

unlimited associative learning (UAL) 178, 187–89
unnaturalness, implications of 123, 127–28, 137–38, 147
unpredictability in behavior 59
upbringing of animals and artificiality to experiments 132

values in science (Douglas 2000) 43n.15, 66–67
valuing putative analgesics/ anesthetics when injured 184–85
vertical homologies 115
vision science, small-N design in 163–64
Visscher, P. K. 142
visual cues 138–39, 182
visual perspective 19
Voudouris, K. 159–60, 204–5

waggle dance 126–27, 142–44
Wagner, G. P. 100n.6
Waytz, A. 58, 59, 60n.12, 62–63
Wegner, D. M. 58–59
Wemelsfelder, F. 196
Wenzel, J. W. 100
whales 98–99, 145–46
Wiegman, I. 105, 106–7
Woodruff, G. 15, 35
working memory 169
wound tending in cephalopods 184
Wray, M. K. 126–27

Yarkoni, T. 74, 204–5
y Cajal, Santiago Ramón 71
Young, A. W. 64–65

Złotowski, J. 63